PROPERTY DEVELOPMENT

PROPERTY DEVELOPMENT

by

A.F. MILLINGTON

BSc(Estate Management), FRICS, IRRV,FIMgt

One time: Professor of Land Economy, University of Western Sydney, New South Wales, Australia; Dean of the Faculty of Business & Land Economy, University of Western Sydney; Professor of Land Economics and Head of the Department of Land Economics, University of Paisley, Scotland; Lecturer in Land Management at the University of Reading; Lecturer in Valuation at the College of Estate Management, University of London. Fellow of the Australian Institute of Valuers and Land Economists, 1988–1996.

2000

A division of Reed Business Information
ESTATES GAZETTE
1 PROCTER STREET, LONDON W1CV 6EU

ISBN 0 7282 0345 6

Reprinted 2002, 2004, 2005, 2007

Cover design by Ted Masters with
photographs from The Stock Market Picture Library,
Andrew Campbell and Andy Tryner

Printed by WS Bookwell Ltd, Finland

market research should proceed forthwith. The objective of such research will be to obtain more reliable information regarding the actual market demand for property of the type which is proposed, and it should seek to determine:

(i) if effective demand actually exists;
(ii) the strength of any such demand;
(iii) the type and quality of development most likely to satisfy that demand; and
(iv) the price limit up to which demand remains effective.

In the past many successful developments were carried out with little or no market research having been undertaken; the developer proceded on the strength of assurances of the existence of market demand from real estate agents and on the basis of natural instinct and optimism. On the other hand, it is possible that many unsuccessful developments might have been avoided if careful market research had been undertaken. Moreover, in today's highly competitive markets it is suggested that market research is essential, particularly in view of the incredible costs involved in many modern development projects. Well undertaken market research should give a very good indication of the strength of demand and of the ability of potential lessees or potential purchasers to actually pay the prices necessary to make a project financially successful. Obtaining more reliable information on such matters than originally possessed is an essential stage in getting a more reliable development valuation than that provided by the initial appraisal.

●Whilst such a stage of the development process does not exist in all development control areas, it is sometimes possible to submit an *outline planning application*. As soon as a developer has decided that a scheme is likely to produce an acceptable profit such an application should be submitted. It enables the planning situation to be clarified, and therefore helps greater certainty to be determined. An applicant will submit an outline planning application which contains the fundamental information relating to the site to be developed and the locality in which it is situated, and the type of development proposed for the site, that is residential, industrial, retail, commercial, or other. It is likely that the provision of information regarding access to the site and the proposed density of development will be required at this stage, but even so the amount of information required will not require the drawing up of very costly plans.

As a result, the procedure enables a developer to seek an assurance that development of the type envisaged will be permitted on the proposed site before going to further expense on the drawing up of detailed plans and on further investigative work. This is helpful to all as it enables the planning authority to consider proposals at an early stage and to indicate its own views about the proposal and its likely requirements with respect to a detailed proposal should the outline application be approved. It is suggested that even when there is no formal process of this type in a development control system, it is wise for a developer to take an informal proposal of this type to the relevant planning authority or authorities to enable the key aspects of a proposed project to be considered, and for discussions and negotiations to be entered into as appropriate. As indicated earlier, the type and amount of development which will be permitted on a site and the conditions which are likely to be imposed on any approval are most important to all parties and particularly to the developer, so the earlier such matters can be considered by a planning authority the better.

Where planning guidelines have been issued in respect of a type of development or an area, a developer should obtain details at the earliest possible time. In any submissions to a planning authority the developer should only depart from those guidelines for good reasons which will need to be fully explained to the planning authority if the proposals are to have any hope of being approved. Strict adherance to guidelines is likely to be essential where such matters as heritage considerations, conservation areas and areas of architectural or historic merit are involved, and in areas of great natural beauty. In general, the provision of guidelines gives greater certainty in an otherwise uncertain situation, and their existence is likely to carry with them the assumption that their observance will normally be enforced by the planning authority.

The *arrangement of finance* is a critical part of any development scheme, and unless appropriate arrangements can be made a scheme is unlikely to proceed. For this reason at an early stage a developer has to have some certainty that required funds will be available. Many developers will have firm relationships with financial institutions which will ensure that, in most cases, funds will be made available to them for schemes on terms indicated to the developer at the time the general arrangement is discussed. Indeed, an understanding of such terms is needed if the initial financial appraisal is to be done with any accuracy. No developer will wish to be in the position of doing a considerable amount of costly investigative work only to find that suitable finance cannot be obtained.

As details of a proposal are firmed up the developer will need to liaise more closely with the financiers and will also wish to get more definite details from them as to the precise terms of a likely loan, as variations in loan terms may have an important effect on the potential profitability of a project. Where the developer is itself a financial institution, or if a financial institution is involved in a joint scheme with a developer, the problems associated with arranging finance are not likely to be so great, although there will still be strict guidelines to be observed and financial targets to be achieved, even though the availability of finance may be virtually guaranteed for good projects.

The first need for finance may well be for land purchase, and the developer will need to ensure that adequate money will be available for each stage of development as it proceeds. Although the immediate need of a developer will be for "short-term finance" – that is the funds to enable the development project to be undertaken – the wise developer will also seek to ensure at an early stage that the availability of "long-term finance" is also assured. This will enable the developer to retain the property after development if the developer wishes to retain it for investment purposes, or alternatively make it likely that another organization will ultimately be able to purchase the completed development from the developer.

As more and more details become established, such as the cost of borrowing money and the other terms on which money will be lent, and an indication is given by the planning authority of the type of development it is likely to approve, a *detailed financial appraisal* can be drawn up. The developer will by then be in a position to estimate with some degree of certainty the amount of net usable space which can actually be created, and market research should have given an acceptable indication of the demand for such space and its likely value in the market. This will enable the projected value of the development to be more reliably predicted, while knowledge of the cost of finance will in turn enable some of the costs of the development to be more accurately estimated.

The development of a more detailed financial appraisal than the initial appraisal will enable the likely financial success, or otherwise, of a concept to be more reliably predicted and will confirm whether it should be further pursued or abandoned. The detailed appraisal should not be a static appraisal. It should be regularly updated by the developer as more information is obtained and as greater certainty can be attached to the range of variables in the appraisal. It should be used by a developer to constantly monitor a project for the viability of a concept can change with the passage of time. What appeared to be a desirable scheme at the outset may, with the passage of time, become one which should either be abandoned or modified as underlying economic factors may alter with changes in such things as rental values and investment yields, which may alter quite dramatically over relatively short periods of time. Such changes might in fact make a project even more desirable calling for, if planning controls permit, an increase in the size of a scheme. If, however, the developer is not regularly updating the financial appraisal such facts may not be recognized and desirable changes to the scheme will not be incorporated.

To be able to prepare an acceptable financial appraisal a developer will need to have good knowledge of:

(i)   the type of development to be undertaken in terms of the market sector at which it is aimed, its size, and the fundamental design criteria;

(ii)   current market values in terms of both rental and capital values of the type of development proposed, and also in respect of undeveloped land suitable for such development;

(iii) market yields for the type of property proposed;

(iv)   the availability and cost of finance, both for short-term and long-term borrowing, for the latter is likely to be an important influence in determining the eventual market value of a property and the ease of marketing the completed development;

(v)   the type of construction methods likely to be needed for the project and building regulations relevant to such a development;

(vi)   the cost of constructing an appropriate building;

(vii) the type of professional consultants which will be required to implement the scheme, and the likely cost of hiring such expertise;

(viii) the strength of market demand and the likely letting or sale period appropriate for marketing the completed development;

(ix)   the likely costs involved in selling and/or letting the completed development, such costs to include agents' fees and expenses, solicitors' fees and expenses, other legal costs such as stamp duty, advertising costs and promotional costs;

(x)   planning law and planning control policies for the relevant locality;

(xi)   the likely impact on the project of environmental control policies and planning gain policies (eg what financial contribution may the developer be asked to make by a local or planning authority before development approval will be given?);

(xii)   the time periods likely to be needed for:

    (a)   completion of the purchase of a suitable site;

    (b)   designing the development and obtaining planning and building approvals;

    (c)   actually constructing the development, and

    (d)   letting and/or selling the completed development.

A general, but extremely important consideration, which is likely to be in the mind of the wise developer at an early stage of a project, is the *taxation implications of the proposed development*. The impact of taxes on a scheme could change what appears to be a financially attractive proposition into a less attractive proposition if the burden of taxation is likely to be too great. Indeed, tax considerations can result in otherwise attractive development propositions not proceeding, and there are many who believe that this has been the case in circumstances in which governments have sought to impose high levels of "development tax" on what they have often regarded as the inordinately high profits made by property developers and the owners of land suitable for development. The wise developer will consequently consider the likely impact of taxation on a project at an early stage in the development process.

The *design stage* of a project is likely to determine the factors referred to in (i) above and is likely to begin very soon after the development concept is formed through the formation of a "development brief". This will indicate, initially in broad terms, the objectives of the development which might, for example, be to develop 5,000 square metres of office accommodation to satisfy demand in the secondary market of a specific suburb. This broad brief is likely to be refined as other stages of the project are pursued, and in due course a "design brief" will be drawn up which will indicate to the architect, in quite specific terms, the requirements of the design exercise and the considerations he or she has to take into account in actually designing a specific building to satisfy both the development brief and the design brief. The design brief cannot really be finalised until a specific site has been secured (but not necessarily actually purchased – see "site purchase stage" later), as detailed design considerations will be very much affected by the location and physical features of the actual site. While the architect will be the main person responsible for fulfilling the design brief, it is likely that inputs will be required from other members of the development team also, and quite possibly from the end users and the end purchaser if there have been pre-commitments to the leasing of accommodation by lessees and to a purchase by an investor.

In development control areas in which a two-stage process of planning approval exists, that is an outline approval stage and a detailed approval stage, the developer will proceed to the submission of a *detailed planning application*. In circumstances in which planning approval must be obtained before development can occur (and where planning control exists that will be in most circumstances), this must be obtained before legal development can proceed. Without development approval land is likely to remain usable for only its current use. Detailed approval determines such matters as the precise amount of accommodation which can actually be constructed, in addition to such matters as details of access to the site, road layout, design features, the type of materials to be used in construction, and other considerations such as landscaping requirements. Accordingly, obtaining detailed planning approval enables the developer to determine with greater accuracy both the likely value of the completed development and many of the costs of creating that development.

If at this stage the detailed planning approval looks more onerous than either the outline approval or the informal indications provided earlier by the planning authority inferred, the financial appraisal should be updated as a result of which the implications of the more onerous conditions should become clear in money terms.

Such matters should become even more clear when a submission is made for *building control approval* as the conditions imposed by those responsible for such approvals are likely to crystallise further the cost of construction of the project, so enabling even more refinement of the financial appraisal to occur. This stage of the development process will require the drawing up of detailed plans for the project, these being sufficiently detailed to allow those responsible for building control to interpret exactly what type of development is proposed and the type of construction methods and materials which will be used. Those ultimately responsible for the actual construction will also be able to build directly from those plans. The preparation of such plans is a complex and costly process for even relatively simple schemes, and for this reason the developer is likely to leave this stage until very late in the development process when as many as possible of the other items previously discussed have reached a stage of certainty. For instance, a developer will certainly not wish to incur the cost of the preparation of such plans only to discover that he cannot obtain loan funds to implement the project.

Once building control approval has been obtained the developer can arrange for the *costing of the construction*, probably by a quantity surveyor, and the preparation of *tender documents and a construction contract*. The costing by a quantity surveyor should allow even more certainty with respect to building costs to be built into the financial appraisal, whilst the subsequent issue of tender documents and the placing of a construction contract should further refine the accuracy of cost predictions. Later in this book consideration will be given to some of the types of contract which may be used and to the selection of an appropriate builder.

Site purchase is a most important stage of the development process. The acquisition of a suitable site at an appropriate price can greatly increase the likelihood of a development being financially successful, whilst the acquisition of a less than ideal site, or payment of too high a sum for a site, may reduce the likelihood of such success. Consequently, at a very early stage of the development concept, a developer is likely to start a serious search for the best possible site, and he or she will hope to clearly identify a suitable site early in the process. However, the acquisition of a site will, in most development projects, be a very expensive part of the development process, whilst, once a site has been acquired, holding it will also be expensive as, until the development is actually completed, there is likely to be no income produced from the site. In addition interest charges will be incurred on the sum paid for its purchase and the other costs incurred when purchasing it.

It is therefore wise policy for a developer, once the appropriate site has been identified, to seek to ensure that, should all other aspects of the development ultimately indicate that it should proceed, the site can in fact be purchased at an acceptable price, but to seek to delay that purchase until the latest possible moment in order to restrict the costs of development. Ideally, site purchase should not occur until actual construction is ready to commence, and there have been cases in which developers have actually managed to defer site purchase until development has in fact been completed. Such tactics reduce the interest charges incurred, but the developer has to ensure that he or she is not put in a position in which the site owner can drive a hard bargain with respect to the purchase price at a stage in the process when the developer's bargaining position has been weakened because of previous commitments in terms of the time and money already spent on the project.

A developer will therefore seek to identify a site at a very early stage of the development process, such a site being suitable both in terms of location and in the ability to physically enable appropriate construction to take place. The developer should seek to agree a price and other terms of purchase with the current site owner and to ensure that the site can be purchased if the project eventually proceeds, using either an "option to purchase" or a "conditional contract" for this purpose. With an option to purchase the developer will pay the site owner a relatively nominal figure "up-front" for the right to buy the site by an agreed later date for an agreed sum of money. If a conditional contract is used, the contract should specify the eventual purchase date and the eventual purchase price, but it should also stipulate that the contract only becomes operative on the occurrence of certain events, which would almost certainly include the obtaining of a suitable planning permisson to allow development to proceed.

The objective of such procedures is to ensure that the site can eventually be purchased on agreed terms but that the developer does not actually have to commit to the expense of a purchase until all other matters considered above have proved favourable to development. Once a site has been acquired the developer will be in a position to place a building contract, but that should not be done until the site has been purchased or otherwise secured.

Supervision of a development is a very important process as without good supervision otherwise sound schemes can become failures. Initially, all supervision is likely to be done by the developer who comes up with the development concept, and supervision may indeed remain the responsibility of the developer throughout an entire project. However, with large schemes a developer is likely to delegate responsibility for some or all aspects of a project to others, possibly setting up a development team with a project manager responsible for its efficient operation.

Whatever the arrangements may be for supervision, the objectives will be to ensure that a scheme proceeds to an acceptable timetable, that the development is to an acceptable standard of design and of an acceptable quality, and that it is completed within approved cost limits. Shortcomings in supervision can be very expensive in that costs may be increased or completion may be delayed with a resultant delay in the receipt of income from the development. Additionally, if the standard of design and/or construction is deficient it may be difficult to dispose of a development, or the proceeds of disposal may be reduced. Supervision of a development will be considered further in a later chapter.

The *marketing of a project* can in fact make or mar its success, and good marketing of an average scheme could result in it being more successful financially than a better scheme which is poorly marketed. Although this topic as a stage of the development process has been left until last, marketing can, and arguably should, be commenced very early in the development process. Indeed, market research and marketing are very closely related, as thorough market research may in fact identify, at a very early stage of the overall process, those to whom a project should eventually be marketed. Nowadays many projects, particularly those involving big financial commitments, will not be begun unless they have, at least in part, been successfully marketed before the scheme is commenced. So major retail developments or major office developments are unlikely to be started until substantial amounts of space have been "pre-let", and commencement of such schemes is also likely to be dependent upon a purchaser of the completed development also having entered into a contract to purchase. In order for this to occur marketing must in fact begin at the development concept stage and if lessees and purchasers can be found at this stage the overall risk attached to a project is considerably reduced. Such risk reduction may be at the cost of lower rentals and a lower capital value being agreed with those committing themselves, for in such cases some of the risk of a scheme is transferred to the committed lessees and end purchaser, in return for which they are only likely to commit themselves "if the price is right", that is if they get a discount for committing themselves in advance of the project starting. In reducing risk by such tactics, a developer is almost inevitably obliged to sacrifice some potential returns, whilst if rental and capital value increases occur in the market during the development period the developer will not benefit from them, as rents and a sale price will have been agreed previously at lower figures.

The successful planning and completion of a development scheme is dependent upon a developer retaining close control over every stage of the development process, and each of the stages referred to above should be undertaken thoroughly and with great attention to detail. Circumstances will in each case determine how accurately each stage can be undertaken and how reliable the results will be, whilst the circumstances of a particular project will also determine the precise order in which each stage will be tackled. This is likely to vary from project to project, but whatever the project may be, it is likely that a number of stages will in fact be undertaken simultaneously, particularly if different staff or consultants are involved in different tasks. The roles and objectives of various specialists will be considered further in the next three chapters.

Chapter 3

Property Developers

Whilst the person or organization commonly referred to as the "developer" is the key person in a development project, the developer will in reality regularly recruit many others to assist in a project, frequently forming a formal development team to take responsibility for a scheme, particularly when it is large and when it involves substantial sums of money.

Those involved in a project can be conveniently separated into two categories, namely:

(i)　developers, and
(ii)　the professionals, or the consultants, who assist and advise developers.

In the second category there is a wide range of professional skills which may be needed by developers, and the consultants employed might include property valuers, real estate agents, planners, architects, quantity surveyors, a range of engineers including site and structural engineers, elevator consultants, heating and ventilation consultants, interior design and furnishing specialists, catering advisers, information technology consultants, landscape architects, marketing consultants, accountants and financial advisers, tax consultants, and solicitors. The roles of these consultants will be considered in a later chapter, this chapter concentrating on consideration of property developers.

"Entrepreneurs" are frequently referred to in the commercial world, and in property development the words "developer" and "entrepreneur" are regularly used as if they are synonymous, and it is worth considering exactly what is meant by the word "entrepreneur" in particular. *The Concise Oxford Dictionary* defines an entrepreneur as the "person in effective control of commercial undertaking". *Everyman's Dictionary of Economics* (J.M. Dent & Sons Ltd, London, 1965) describes an entrepreneur as an

> enterpriser; one who ventures on or undertakes an enterprise; sometimes applied to a firm or more generally to entrepreneurship as decision-taking. How far business men are individual entrepreneurs in

the sense that they risk their capital and own their businesses is not known. Whether the present-day top executive should be called an entrepreneur is debatable . . . In the large public company, however, the board of directors and the senior executives who take the major decisions are not risk-bearers; they do not venture their capital in their own business. The ownership by shareholders is largely separated from executive control.

So, as early as 1965 the concept that those in charge of property developments were automatically risk-takers was questioned.

These observations are even more apposite to the current property development scene in which many large developments are now undertaken by large commercial organizations rather than by individual entrepreneurs risking their own money, the decisions to develop being made by committees of which few, if any, members may actually have "risk money" involved in a project. The only risk carried by many of these decision makers is that, in the event of a project with which they are involved being unsuccessful, they may lose their seat on the Board of Directors or their job with the organization if they are an employee. Even those are possibilities which, it would appear from the way most large organizations operate, are only likely to be implemented in the event of a succession of failures rather than in the event of one failure, even a particularly bad one. Those making the decision to develop therefore rarely carry the big risks traditionally associated with property development, those who stand the risk of losing their money being shareholders or unit trust holders who in general have no say in the decision making process. Their only part in such a process is very indirect, being their right to vote at annual general meetings or special general meetings if they are held.

The expression "developer" is defined in *The Glossary of Property Terms*, (The Estates Gazette Limited, London, 1989) as

> An entrepreneur who has an interest in a property, initiates its development and ensures that this is carried out (for occupation, investment or dealing) and from the outset accepts the ultimate responsibility for providing or procuring the funds needed to finance the whole project.

This definition aptly describes the role of the developer with the exception that one could quibble with the use of the word "entrepreneur". However, the definition infers in the last ten words that the developer may not necessarily be risking his or her own money, and in that respect the modern property developer is perhaps aptly defined.

The developer is in essence the person who forms the development concept, who initiates the project, and who remains responsible for financial aspects of the project even though his or her own money may, in the modern development scene, not be at risk. While the responsibility of the developer may in such circumstances be very large indeed, particularly with major developments, there is nevertheless the likelihood that the prospect of possible financial failure is not such a daunting prospect as it would be if the developer's own money was at risk. Indeed, it is likely that in recent years some decisions have been made to develop in circumstances in which an individual developer would not have risked his or her own money.

Developers may fall into different categories with different development objectives, although they will all certainly have the common objective of wishing to complete successful developments, and the type and form of development is likely to vary depending upon which classification a developer falls into.

Those who build for their own use will have the prime objective of ensuring that the development is ideally suited to their own needs, but even in those circumstances they may be wise to ensure that the design of the property they develop is such that it is likely to have good market appeal and good market value should they in due course decide they no longer need it for their own use. Such developers will include companies which build for their own operational purposes and individuals who build houses for their own occupation and, while their activities may form a relatively small proportion of the total development scene, they are nevertheless active players in the market.

In deciding whether to build exactly for their own needs or whether to consider the wider market appeal of their developments they are likely to be very much influenced by the extent to which compromise of design is likely to adversely affect their own operational or user efficiency. Where the effect is likely to be slight they may well design for the wider market, but where substantial adverse effects are likely to result they are unlikely to compromise their user requirements. As a result, although such a property is likely to have very limited market appeal and consequently a low general market value, a brewer is likely to design a building precisely to satisfy brewing requirements as thereby the efficiency of the brewing operation and the profits to be made from the same are likely to be maximized. In such circumstances the market value for a possible sale some years hence may be only a remote consideration for the developer.

Those who build to let to others are likely to be very much influenced by the type of property demanded by the majority of those who rent property, so, depending on specific market circumstances, they might decide to build to a very high standard of design and construction, or alternatively to build to a more basic level of design and construction if those wishing to rent property only have limited resources. They are likely to do exhaustive market research to determine what the market wants for what could be described as *"demand-led development"*, that is the development occurs as a result of the identification of actual unsatisfied demand in the market.

Those who build to let to others may also be involved in more speculative types of development where it is not necessarily possible to identify the existence of unsatisfied demand in the market place. It might be argued that to develop in such circumstances would be unwise, but it should not be overlooked that property development organizations are set up to develop property, and if an organization has a team of staff and fixed capital it is likely to actually lose money if no development is undertaken. With fixed outgoings which cannot be avoided, the tendency will be for a company to continue to develop even when demand is limited, but to seek to develop in a way which reduces financial commitments as long as such a policy does not thereby reduce the chance of likely success. In such circumstances the developer will probably try to anticipate what "the average person in the market" (if such exists) would wish to rent or buy, and to build such properties with, however, strict cost limits in an attempt to control the possible "down-side" of a project. Such development can be described as *"supply-led development"*, being undertaken primarily on the initiative of the developer who will attempt to "hedge his or her bets" by building property with as wide an appeal as possible. Market research is likely to be important in such circumstances in helping to determine exactly what type of property is likely to have wide appeal.

In distinguishing between demand-led and supply-led development, it is appropriate to point out that the former is likely to be based on the supply of property for specifically identified tenants or tenant groups, whereas supply-led development will be based on construction to suit only possible tenants who have not as yet been identified as being in the market-place. With demand-led development there is therefore the likelihood that specific needs have been expressed to the developer who will be likely to pay particular attention to such things as locational needs and to what may be quite rigorous user requirements.

Those who build to sell for profit, rather than to retain and let out properties, are naturally hoping to make the maximum possible capital profit from an immediate sale, and they may be influenced much more by short-term considerations rather than long-term considerations, the latter being more important to those who retain properties as investments. Developers in this category are likely to be very much aware of the current needs and purchasing ability of those who buy property investments if they are developing properties for office, retail or industrial use, and will accordingly build to satisfy investor needs. Those who are building residential properties for sale will seek to develop properties which can be afforded by and which are likely to satisfy the user needs of identified market sectors. So the developer will seek to determine whether those needing houses are primarily, for instance, first-home buyers with limited funds, or those in the middle-aged executive class with high incomes, for each class of potential purchaser is likely to require a very different type of property. Identification of market need will be important for such developers, as the objective will not only be to maximize the profit made from each property, but also to sell each property rapidly in order that the returns and profits made may be used to fund further developments.

Those who build to create investments for sale to others will have to consider not only the requirements of potential lessees but also the needs of long-term investors. They will undertake a development and will let it before selling it to the long-term investor and, if they are to attract a long-term investor, they will have to ensure that the design and quality of construction of the development, the type of tenants to which the property is let, and the terms on which leases are agreed, are all likely to be acceptable to an investor. Such a developer may be well advised to try to identify an actual investor-purchaser before commencing a scheme, and to design the actual scheme as closely as possible to satisfy the investor's needs.

Public bodies which develop to satisfy their own operational requirements and social needs may have quite different considerations to take into account. Their major objectives will be to satisfy those requirements or needs, and the profit motive may not enter into the equation, although they are certainly likely to have to develop in a way which ensures that their costs are contained within an acceptable budget.

In some cases development may be undertaken by public authorities in an attempt to increase employment in an area and, whilst the profit element of such a venture may be the reduction of the social costs of unemployment, they will need to operate very much in the same way as commercial developers operate with respect to seeking to identify and satisfy normal market demand if employment is indeed to be increased. Developments such as "starter-factories" and development area industrial estates may be undertaken by public authorities who should, in theory at least, be able to develop them more cheaply, as instead of being constrained by the need to make a profit, their constraint would be the need not to make a loss. However, it may well be that the pressure to make profits results in commercial developers being more efficient developers, and there are probably many examples of public developments of this nature being costly failures because the developers have not in fact acquired market experience "the hard way", through experience in the hard commercial reality of property markets.

A significant change has occurred in the development scene in many countries, since about 1960, with the advent of *institutional developers*. Prior to that time, most developers were private entrepreneurs who borrowed funds from large institutions such as insurance companies. Developers were able to borrow cheaply and the yields obtainable from property developments generally exceeded the cost of borrowing, but with the passage of time many of the funding institutions realised that large profits were being made from the funds they lent to developers, with the result that many of these lending institutions decided to become property developers themselves. By doing this they carry out developments using their own funds at a low opportunity cost, and in turn they receive both the development profit and the investment profit, assuming they are efficient developers and actually make those profits. Since first entering the development scenario many institutional developers have in fact become major players in the property development arena, and many of today's biggest developments can in fact only be contemplated by such developers because of the huge sums of money required for their completion.

Over a similar time period the concept of *joint developers* has also arisen. This expression refers to the type of development arrangement rather than the type of property developed. A typical joint development agreement is likely to be between the owner of the development site and the developer, the two agreeing to jointly fund the development and to share the eventual profits in a pre-arranged manner. The site owner's funding is likely to be through the provision of the site at an agreed value, the developer providing the development expertise and arranging the provision of other development funds required, although it is quite possible that the site owner might also provide some of the additonal funds required if in a position so to do. Such an agreement is likely to arise in circumstances in which site owners wish to participate in possible development profits but do not have the development expertise to realize such profits themselves. Agreements of this nature frequently occur with land owned by public authorities, the public authority land owner and the property developer becoming joint developers. Such an arrangement is generally particularly appropriate for a public authority which has the responsibility of maximizing the returns to property in public ownership, but which is not in general terms set up to operate in an entrepreneurial way, particularly with respect to accepting the risks inherent in property development, and which also is unlikely to possess the personnel with appropriate skills to undertake property development. From the property developer's point of view, while the sharing of profits may not necessarily be attractive, the developer is generally relieved of the need to raise funding for the land, which land might in any event not even become available for development in the absence of a joint scheme. Such arrangements are generally made, therefore, to the benefit of both parties to the agreement.

Whilst the above refers to typical circumstances in which there may be joint developers, such arrangements may arise in other circumstances also. Construction of the Channel Tunnel under the English Channel between England and France was so large an engineering and financial project that it was undertaken by a joint development team comprising a consortium of a number of engineering and financial organizations from both France and

England, such an arrangement being the only way in which so large a development scheme was ever likely to be implemented.

The motivation of property developers

Property developers are business people; just as other business people supply goods and services, so do property developers supply them in the form of completed property developments. They take undeveloped land, or existing developments which have reached a stage of economic obsolescence, and using their entrepreneurial skills they develop or redevelop the land in a way which they believe will attract potential purchasers to the finished product.

Any success achieved will depend upon their abilities in a range of activities; the ability to find land suitable for development or redevelopment, to assemble land into viable development parcels; to envisage appropriate forms of development for sites (that is to have vision); to accurately predict market demand; to plan schemes; to arrange suitable finance for schemes; to manage developments; and to market them satisfactorily. They will only succeed if they develop land in a manner that will satisfy market demand and so result in sales at prices which produce an appropriate profit for them.

The philosophy of developers

The basic philosophy of developers is that development propositions have to be financially viable otherwise they will wish to have nothing to do with them. Developers cannot afford to become involved with propositions which are likely to result in losses; they carry out property development as a business venture, and a successful business must make profits. In addition, those profits must provide an adequate return on the capital employed in any venture plus an additional return to reflect the level of risk involved. In particular, a development needs to provide an adequate return on the developer's equity (the developer's own money invested in the venture) if it is to be considered a success from the developer's point of view. If a venture is unlikely to provide an adequate return to the developer, then alternative projects or investments should be sought and entered into.

The qualities needed by property developers

If they are to be successful, property developers need a wide range of personal qualities which can be easily listed but which include such a range of qualities that they are rarely all found in any one individual. Reference to many of these qualities and consideration of them in greater depth will occur in later stages of this book, and they include:

(i) optimism;
(ii) imagination and practical vision;
(iii) a thorough knowledge and understanding of the market for property;
(iv) a thorough knowledge and understanding of the construction process;
(v) an understanding of finance and the ability to raise money in the market: developers must be able to inspire confidence in potential lenders;
(vi) the ability to manage development schemes and in particular to be able to respond to changing circumstances which are inevitable with the passage of time;
(vii) the ability to make judgements;
(viii) the ability to make decisions and to act upon them;
(ix) "a thick skin" , that is the ability to withstand criticism; and
(x) courage.

Two prominent developers who heard me give the above list of qualities during a conference afterwards told me I had omitted only one thing – "a sense of humour", they told me, is another essential quality for any developer who hopes to survive in the competitive commercial environment.

The major development role of developers

Further consideration of the basic development equation enables it to be rearranged to indicate the profit which any project should produce:

			Value of the completed development
Less		(1)	Costs of development
	plus	(2)	Cost of site purchase
	plus	(3)	Costs of marketing the development
	plus	(4)	Cost of financing the development
	Equals		The development profit

It can readily be seen from the equation in this format that anything which either decreases returns or increases costs is likely to threaten the financial stability of any development scheme, as each will reduce the prospect of adequate profits being made. Any likelihood of such changes occurring will deter developers, and they know that some will inevitably occur through "the sheer cussedness of life" as Jerome K. Jerome called it in *Three Men in a Boat*. Developers hope that some movements when they occur will be in their favour, for example an increase in the market value of houses which would increase the returns to a residential development, or a drop in the rate of interest charged on borrowed money which would reduce the costs of financing a scheme, and which might also reduce the costs of potential purchasers thereby making them better able to purchase the finished product.

However, the stronger the probability of factors moving against them, the more likely property developers will be not to get involved with a particular project. For instance, they will be reluctant to develop houses in areas of high unemployment, reluctant to develop factories in areas of low productivity with resultant doubtful rental returns, and reluctant to develop anything in areas where unacceptable planning delays and other planning problems are likely to arise. There were periods in the 1970s when developers were loathe to have any involvement at all in certain cities in the United Kingdom in which development was desperately needed in terms of economic, physical and social considerations, such reluctance resulting from the fact that political attitudes (and resultant planning policies) in those cities were very much "anti-developer". Consequently, despite the fact that the skills and enterprise of developers were greatly needed in those cities, developers did not want to become active in those areas because of the increased level of risk resulting from local political and planning policies, such policies threatening their prospects of running profitable development projects.

As already indicated anticipated profits may not in fact be realized for one, or both, of two main reasons:

(i)   there is a shortfall in expected returns or
(ii)   there is an increase in the forecast costs of development.

A shortfall in expected returns may result from lower demand for the completed products than was anticipated, which may be caused by a variety of factors including:

(a) miscalculation or misjudgement on the part of the developer;
(b) an increase in the rate of interest payable on borrowed money which reduces the ability of purchasers to buy the product;
(c) a shortage of loan funds which has the same result as (b) and which may occur as a result of changes in bank policy or government policy;
(d) competition from other developments;
(e) recession in the local economy which takes purchasing power away from potential users and purchasers of the property;
(f) a restriction in the density of development allowed on the site by the local planning authority which reduces the amount of property which can be developed and sold.

An increase in the costs of development of a project may be caused by such factors as:

(1) an increase in building costs which may be caused by, for example, unexpected problems in working a site such as the discovery of unstable ground, underground water, old drains, or inadequate natural site drainage characteristics;
(2) an increase in the rates of pay of building workers;
(3) an increase in the overheads associated with the employment of workers, such as increased superannuation or insurance contributions, or the imposition of or increases in employment taxes;
(4) an increase in the cost of building materials;
(5) an increase in the rate of interest charged on borrowed money which increases the costs of financing the development; or
(6) an extension of the building period which causes increased financing costs and which may result from such things as inclement weather, industrial disputes, a shortage of building materials, planning problems, or problems with respect to building regulations.

A major task for property developers is to seek to mitigate the effects of such potential unfavourable occurrences, and, if possible, to completely eliminate the chance of them occurring. There are a number of appropriate steps which can be taken to achieve such objectives, and these risk control measures will be considered at greater length later in this book and in Chapter 17 in particular.

The role of property developers in society

In general property developers do not have a good public image
and are much maligned people, in many cases, if not most cases,
undeservedly so. The public image of them is often of people who
make large and unjustified profits from their activities, and the
high-living profile of a minority of developers has probably given
rise to such an image. In general their deficiencies and
shortcomings and the misdeeds of the few get very prominent
publicity in the media, whilst, on the other hand, it is rare for a
property developer to get unsolicited and favourable publicity.

It is therefore not unreasonable to question whether property
developers have a valuable and acceptable role to perform for
society. We live in an age of specialism, and in the modern world
there are few generalists who tend to all their own daily needs and
who produce all they need for their own and their families'
survival. Such people are generally only found in Third World
countries and in areas where modern civilization has as yet barely
touched. We are nearly all of us dependent on a range of goods and
services produced by specialists. In the same way as cars are
produced by specialist car manufacturers, who themselves employ
specialists in the range of skills they need to produce their cars, and
foods are produced by specialist stock rearers and specialist
growers, so do property developers produce the types of property
that are required by society in general for a wide range of everyday
activities.

People have a need for property for virtually every activity they
perform in everyday life, and so there is a total need for specialized
properties to live in, to work in (be the work commercial,
industrial, retail or another form of employment), to be educated
in, to worship in, for recreational purposes, for religious purposes,
and for a wide range of other activities also. As society develops
with higher expectations, the types of buildings demanded for
these various activities become more sophisticated, whilst the laws
which regulate the construction of buildings tend to become more
complex as developing society demands ever higher standards of
performance, comfort, health and safety.

The property developer performs a valuable role for society in
just the same way as do other specialist providers of goods and
services. Developers seek to analyse market needs and to provide
suitable properties to satisfy the needs of people. Developers seek
to be successful through providing suitable properties, in

appropriate locations, at appropriate times to satisfy demand when it arises, of the right design, of an acceptable quality, and at an appropriate price to enable those who need or want the properties to be able to rent or buy them.

In doing this developers enable others in society to concentrate their minds and energies on their own specialisms. Developers remove from others the problems of providing suitable properties, and together with investors, who subsequently purchase many developed properties, they also remove the problems of property management from those who are not specialists in property matters. Developers also accept the risks of property provision and in so doing remove those risks from the shoulders of others, so enabling those others to concentrate on risk control in their own specialist areas where they are more competent and therefore more likely to earn success for their efforts. Consequently, business people in particular are free to concentrate their skills and energies on practising their own pursuits, and the total productivity and economic success of society as a whole can thereby be increased.

Collectively property developers and property investors therefore provide a pool of appropriate properties to allow other members of society to pursue the whole range of human activities with greater efficiency. They therefore make contributions which benefit society in general and, by enabling it to function more efficiently (through the provision of functional properties), they assist in the reduction of both waste and the inefficient use of scarce resources. Moreover, by using high levels of skill they can refine the quality of the products they produce for the benefit of all. One only has to compare the efficiency of industrial workers employed in modern industrial properties to those employed in older, less efficient, less healthy and less safe premises, or office workers employed in efficient, modern buildings to those employed in offices built many years ago, to realize how important the provision of efficient properties is to society in general.

Property developers most definitely fulfil a very important role in society. It is probably unfortunate that the excesses of a few high-profile people, and the failures of some often in spectacular circumstances, should detract from the general public image of developers to the disadvantage of the many competent and successful developers who provide substantial benefits for society as a whole.

Chapter 4

The Role of the Property Investor in the Property Development Process

The concept of investment

Property development and activities in the property market revolve around money, and in that respect they are no different from most other activities in life. There are few activities that do not require money, whilst decisions such as whether to buy a car, and if so whether to buy a small car or a large car, or whether to buy a holiday, and if so what type of holiday, all involve a need for money and a judgement on how to spend that money. In the same way the decision to develop property or to buy a property involves the need for money and a conclusion on how to use that money, that is a decision will have to be made whether to use money for investment or not. Both the development and the purchase of a property involve investing money, that is saving it rather than using it for immediate consumption.

The layman would probably describe investment as spending money on articles or securities which will be retained in one's ownership and which are expected to appreciate in value in the future and which hopefully, in most cases, will also produce a periodic income flow. Economists define investment slightly differently. In *Basic Economics* (Macmillan Education Ltd, London, 1981) J. Harvey describes investment as:

> . . . spending over a given period on the production of capital goods (houses, factories, machinery, etc.) or on net additions to stocks . . .

This definition highlights the need to distinguish between the economic concept of investment and putting money in a bank or buying securities, which actions could well come within a layman's concept of investment. In *Everyman's Dictionary of Economics* Arthur Seldon and F.G. Pennance defined investment as:

> Man-made assets which are used in the production of consumption or of further investment goods,

whilst Paul A. Samuelson in *Economics – An Introductory Analysis* (McGraw Hill, New York, 1961) said:

> . . . we must recognise that the final goals of people do include net investment or capital formation, not simply current consumption.

Irrespective of apparent conflicts in the definitions of laymen and economists, they both in fact have many things in common. Implicit in the concept of investment is the sacrifice of immediate consumption, that is money is spent on something which is permanent in nature and which will give returns over a long period of time, rather than on items which are immediately consumed and which therefore give short-term benefits only. If the average person is to sacrifice the benefits of immediate returns they will usually only do so if they receive rewards for so doing. This is basic common sense apart from sound financial sense, and the fact that one can see a sacrifice vividly if one gives up the use of one's own money makes the logic of the proposition even clearer.

Some people suggest that requiring a return for the use of or sacrifice of money is immoral, a view which is quite strongly held by some even in the modern, commercially orientated world, particularly by many who profess to be communists or socialists. Usury, the practice of lending money for interest charged, has frequently been scorned by commentators, and as long ago as 1551 an anonymous quote said "No Christian is an usurer . . ." inferring that Christian beliefs and usury were inconsistent with each other. In the eighteenth century Edward Gibbon referred to "The usurer, who derived from the interest of money a silent and ignominious profit."

However, the money possessed by someone represents goods and services in that its expenditure could result in their purchase. Is it really therefore immoral to charge hire fees for its use, which is what interest charged amounts to, when if that money were exchanged for such things as cars, flats, tools and machinery, the borrowers of such items would fully expect to pay a hire fee for their use and would see nothing unusual or unreasonable in having to do so? They might well consider charges made to be too high, but they would not consider it unreasonable to be expected to pay for the hire of goods and services, this being a perfectly normal commercial arrangement.

Whatever the ethical considerations may be, it is a fact that those who invest funds give up the immediate use of their money, and in doing so investors make savings which, if they are invested in a sensible way, will provide them with returns over time. At the same time, if property investments are made available for people and organizations to lease, they will provide others, who are unable to purchase their own properties, with the opportunity to rent their property investments for occupation and use. In such situations both the investor-owner and the lessee should benefit from the investment decision made by the owner of funds.

For investment to be a sensible decision, a potential investor will expect future returns to be at least as great as those which could be obtained from immediate consumption, otherwise immediate consumption would be more beneficial overall and a more sensible option. However, because future returns will be delayed and because greater uncertainty attaches to the future than to the present, the wise investor will require the anticipated future returns to be sufficiently large to both equal the returns which could be obtained from immediate consumption, and to compensate for the delay in receiving future returns and the greater uncertainty which inevitably attaches to them. The more distant they are and the more uncertainty attached to the future returns, the greater will be the compensation the wise investor will require, and the potential returns from investment therefore have to be sufficiently large to actually persuade people or organizations that investment is sensible, and to compensate them for the overall risks inherent in any type of investment.

Investment opportunities available

Just as it would be unwise to invest unless the returns to be expected from investment will be at least equal to and preferably greater than the returns to be obtained from immediate use of funds (or immediate consumption), so would it be unwise to place money in one form of investment if greater returns can be made from another form of investment without incurring increased risks or unacceptable disadvantages of any kind. A potential investor should therefore always compare the relative qualities of the range of available investment opportunities, at the same time as comparing the qualities of what is decided to be the most acceptable investment opportunity with the benefits (or returns) that could be obtained from spending the funds on immediate consumption.

In making comparisons between different possible investments, the degree of importance attached to the different characteristics of each possible investment will vary from person to person and from organization to organization. This will be because of a range of considerations including the objectives of investment; the size of the funds available; the length of time over which funds can be committed by the investor; the expectations of each investor regarding the future and the future of the economy in particular; their other financial commitments; and their other existing or anticipated investments. Not only will assessments vary from investor to investor, they are also likely to vary with each investor at different points in time, such changes in view resulting from changed personal or company circumstances and from changes in external factors also, such as the state of the economy. With an individual at different stages of life there will be different personal circumstances which result in investment capability and policies varying over time. The young, single man with high earnings and high investment capacity may become a married man with high family commitments and little investment capacity, whereas with the further passage of time he may be in a state of high income, low outgoings and high investment capacity once more. Similar differences may arise between an infant company fighting to become established and to accumulate capital, and a mature company with an established capital base and high investment capability.

The major factors in determining who invests and what they invest in are often the amount of capital required for specific investments and the degree of risk which attaches to each investment. Those with limited amounts of capital available will generally find it difficult to invest in property because even the meanest property investment will normally require a substantial capital investment. If they have limited funds available they are also likely to find the high risks and disadvantages of property investment to be unacceptable. At the end of the day, the type of investment made by an investor is likely to depend upon the size of funds available, the objectives of investment, and the level of risk attached to any investment related to the ability and willingness of the investor to be exposed to specific levels of risk.

The characteristics of property as an investment medium

As suggested above, there are many disadvantages to property as a form of investment when compared with other investment opportunities. It is generally accepted that qualities which make investments attractive are:

(i)   the provision of an adequate and regular financial return to compensate the investor for the loss of the use of the money invested and for the level of risk involved in making the investment;

(ii)   ease of initial investment and low costs of initial investment;

(iii) ease of sale coupled with low sale costs;

(iv)   short marketing periods for both purchase and sale;

(v)   no, or only low, periodic ownership and management costs;

(vi)   divisibility of the investment allowing a partial sale of the investment if necessary.

It is a fact that property as an investment generally possesses few of these attractive features, and compared with the list above the qualities of most property investments are:

(a)   if carefully and well selected, as should be the objective with any investment, property investments should provide regular income flows, through rent paid, sufficient to reward investors for the sacrifice of the immediate use of their money and sufficient to compensate investors for the risks inherent in the particular investment;

(b)   it is not easy to purchase property, professional advice generally being needed and relatively high costs being involved, frequently as high as 5% of the purchase price agreed, and sometimes even higher;

(c)   similarly, the costs of sale are relatively high, whilst sale is only possible if there is a potential buyer, the heterogeneity of property interests regularly resulting in this not being the case in the short-term;

(d)   because of (b) and (c) above the marketing period for a property is generally long in comparison with other investments, whilst proof of ownership is not always easy to establish and may also be a costly process;

(e)   the management of property interests is often complex and consequently generally results in relatively high management costs;

(f)   it is regularly either impossible or impracticable to divide a property into smaller parts for sale purposes, and accordingly if capital recoupment is required it can only be achieved by the sale of the entire interest which may require a lengthy time period and which may also be costly.

Whilst the above suggests that property has many unattractive features as an investment, its ability, when carefully chosen, to produce regular income flows and the fact that over substantial periods of time many, probably most, property interests have proved to be good hedges against inflation in real terms, means that despite the disadvantages it still remains very attractive to many investors. However, its value as an investment is completely dependent upon its utility to would-be users, and in times of recession or when there are changes in underlying economic conditions, it may in fact become a liability rather than an asset. Whilst stocks and shares which produce no income for periods of time are likely to cause worry for their owners, no holding costs are likely to be incurred by the owners in the period in which no income is received. However, with property there are likely to be substantial holding costs even when there is no lessee and therefore no income, such costs including insurance, repairs and maintenance, and rates and taxes. The assumption that property will automatically appreciate in value over time is also false, and where obsolescence occurs, be it because of physical or economic factors, property can in reality depreciate in value over time.

The role of property as a form of investment

A country's people are its most valuable investment, but second in importance as a valuable asset is a country's land and the developments thereon. It is therefore important, if a country is to maximize its economic and social potential, for there to be sufficient and appropriate investment in property development. As indicated earlier, as far as those in possession of finance are concerned the main attractions of property as an investment are the security in real terms of the capital invested in it, and its ability to produce an adequate and regular income flow. These qualities make property particularly attractive as an investment to organizations which are entrusted with other people's funds and which at the same time have the need for regular income flows which are inflation-proof (that is their purchasing power is not

Loan Receipt
Liverpool John Moores University
Library and Student Support

Borrower ID: 21111107157158
Loan Date: 29/10/2009
Loan Time: 7:21 pm

Property development /
31111012117410

Due Date: 19/11/2009 23:59

Please keep your receipt
in case of dispute

likely to be depreciated by the effects of inflation). Consequently, in recent years in particular, major investors in property have included insurance companies and superannuation funds.

Apart from its qualities in investment terms, investment in property also enables such organizations to "balance their portfolios" by having a mix of investments covering such investments as government stock, equities, and property, the latter giving contrast to the other types of investment and so providing greater stability to an investment portfolio. The very fact that property investments have different features to the other investments gives them attractions in helping to reduce the overall risk attaching to a portfolio.

The role of the property investor in the development process

Property developments generally require substantial sums of money and many developers would be unable to undertake developments unless they could reasonably contemplate the purchase of their completed developments by someone else. This is so in part because many developments are made possible through the use of borrowed funds which have to be repaid by the developer out of sale proceeds once a development is completed, whilst many developers also need to realize their own equity in a development once it is completed in order to be able to finance further development activities. The existence of property investors provides the market for the developer's products, and their purchase of completed developments enables continuity of development activity to occur, thus assisting in the development of property to satisfy new economic needs and the redevelopment of property which is no longer needed to ensure that the total building stock is up to date and efficiently used.

In this respect property investors fulfil useful roles in society, but they also assist economic activity by creating a situation in which would-be property users do not have to tie up substantial amounts of finance in property purchase. The availabiltiy of a stock of property owned by investors and available for rental enables property users to pay for property use on a periodic basis and to use their capital for investment in their own business activities. Property investors therefore assist the adequate or more satisfactory financing of general business activity in the economy, as without their existence to fund the long-term provision of suitable properties, much worthwhile business activity might not occur, or might only occur in an under-financed and therefore less efficient way.

Another aspect of this situation is that by making properties available for rental on a periodic basis, property investors also provide greater flexibility in the economy as property users do not have to commit themselves to the long-term occupation of one property through purchasing it. The existence of a pool of rental property enables developing firms to grow and move to bigger properties as and when necessary, whilst appropriate locational changes are also possible for other reasons as necessary. Similarly, more efficient use may be made of property by firms moving to smaller units of accommodation if economic conditions make this advisable or if the development of new technologies enables economies to be made in the use of space. Where firms have had to commit themselves to the outright purchase of properties such decisions may not be realistically possible, or, if possible, they may be costly to implement.

The scale of many modern developments is so large that they would not take place at all unless developers knew there were investors willing to purchase them on completion, and as a result many investors are committed to the purchase of property developments before construction even commences. This removes much of the risk from the developer, but the risk is assumed by the investor who thereby assists the development process to take place. In assuming responsibility for what are often very large risks, investors accordingly expect substantial returns over time as compensation, and it should not be overlooked that there is no guarantee of such returns ever occurring. The whole development process takes place in the hope of anticipated returns occurring: such hopes are not based on certainty but only on well-informed opinion backed up, hopefully, by exhaustive and effective market research.

Even if there is demand when a development is begun, the economic scene may have changed substantially by the date a property is completed. If this is the case and unfavourable conditions exist on completion, the investor may well suffer substantial loss, or, at best, suffer reduced returns from those originally anticipated. Anticipated returns may also be badly hit by the advent on the market of unanticipated competition from rival suppliers, whilst the inherent risks of property development referred to in the previous chapter may also have operated to adversely affect the development and, if the investor has assumed the responsibility for any of those inherent risks, then the quality of the investment will obviously suffer.

Yields from property investment

The general rule that the greater the risk attached to an investment, the greater will be the return an investor hopes to get, applies to property investment as it does to other types of investment. So, the more speculative a property development project is, the higher will be the yield (in terms of percentage return on capital invested) which an investor hopes to receive on completion of the project. Clearly, no specific yield can be guaranteed, and an investor will make his or her own assessment of an appropriate yield in calculating the sum they are prepared to pay to purchase a particular property development either "off the plan" (that is before it is commenced) or on completion. If a project is bought "off the plan" there are likely to be more unknowns than if it has been completed, when its success or otherwise is likely to be more easily determinable. In most cases, therefore, one would expect an investor to require a higher yield from an investment purchased "off the plan" as the investor in such circumstances is likely to be taking responsibility for more areas of risk and uncertainty, including those attached to the construction process and the leasing of the project.

Methods of measuring returns

A developer or an investor may use a variety of methods to assess the attractiveness or otherwise of a particular development project.

The capital profit may be a relevant consideration, particularly if an investor is likely to provide all of the funding for a project, this being the measure of the expected profit when all the costs of development are taken from the anticipated market value of the completed development. So in circumstances in which the market value of a project is expected to be £5 million and the total development costs £4 million, the developer/investor expects a capital profit of £1 million, or 20% of the total capital value. Depending on the circumstances of a developer/investor and the risks associated with a particular project, so will the percentage of capital profit expected from a project vary, and whilst some developments might proceed on the basis of an expectation of 10% capital profit, others would not be contemplated for an anticipated profit of less than 25%. There is no golden rule about percentage expectations: each developer-investor is likely to make a specific assessment of requirements for each specific proposition depending on their own circumstances and the details of the scheme, with the result that some developers-investors may be prepared to undertake schemes for anticipated capital profits which others would find quite unacceptable.

The *development yield* indicates the annual rate of return expected to be received from a development and is found by comparing the anticipated net annual income from a project with the total costs involved in creating the project. So if a project with total anticipated costs of £4 million is expected to produce net annual income of £600,000, the development yield is predicted as being 15% per annum. This is a very practical method of assessing the merits of a development project as it enables the likely rate of return from the project to be compared with the alternative of placing the capital sum required in alternative investments such as long-term bonds or equities, and thereby enables the relative returns and relative risks to be compared.

The development yield should not be confused with the *market yield* of the project which indicates the relationship between the anticipated net annual income produced by the completed project and its anticipated market value. Using the above figures, with an expected market capital value of £5 million and expected net annual income of £600,000, the market yield of a project would be estimated at 12% per annum. This would be an acceptable method of measurement for an investor purchasing a completed development, the yield being assessed in relation to the purchase price of the investment.

The *developer's return* is generally estimated by expressing the capital profit which it is expected the project will produce as a percentage of the total costs involved in creating the development. Again using the same figures, with an expected capital profit of £1 million and total development costs of £4 million, the expected developer's return would be 25%. This is a particularly relevant method of assessment in cases in which a developer or an investor is also a construction organization when the return on development costs is likely to be the most favoured method of comparing one possible project with another.

When an appraisal is done using a discounted cash flow format the *internal rate of return* can be calculated to indicate the likely performance of a development project. The rate of interest at which the net present value of the outflows and the net present value of the inflows are equal will be the rate at which a development earns money, and this concept will be considered further in Chapter 14. The IRR, as it is frequently abbreviated to, also permits the relative earning power of alternative projects to be compared.

Depending on circumstances a developer-investor may prefer one measurement of viability as opposed to the others, but it is likely that most will use more than one measure to determine whether it is wise to become involved with a scheme. It should not be forgotten that, at the outset of a scheme, all these tests are based on predictions which may in the event prove to have been over-optimistic, and it is for this reason that some of the percentages required by developers-investors may seem high to the uninitiated. If returns could be guaranteed there is little doubt that lower yields and capital profits would be utilised than is the case where all figures are predictions and liable to substantial change. Indeed, many of the risk control measures used by developers are aimed at reducing the possibility of change in many of the predicted figures, and when this is possible there is frequently a downward adjustment in the required returns also, as part of the overall risk inherent in property development will have been avoided.

The reverse yield gap

This is an expression which one frequently hears in association with property investment, and it refers to the fact that over many years numerous property investors have been prepared to accept lower yields from property investments than they could have received by investing their money in government bonds or in equities. There has been a gap between the initial yields of say 4 to 6% accepted from certain property investments and the higher initial yields which could have been obtained from the alternative investment, and the reason investors have regularly been prepared to accept lower annual yields from property investments has been their expectation that these lower yields would be more than compensated for by growth in the capital values of their property investments. They have therefore calculated the expected returns from property investment by reference to a combination of the

annual yield from the investment together with the annual
equivalent yield of the anticipated capital growth over a period of
time, and have presumably decided that while initially there would
be a "reverse yield gap" with the property investment, the annual
equivalent return of the expected capital growth would be
sufficient to compensate for that initial gap.

In the past this has regularly proved to have been the case, but it
must not be assumed that it is inevitable that capital growth will
automatically result from investment in property. It is only if the
circumstances are right that capital growth will occur, and in
general terms it will be necessary for there to be increasing demand
for a particular type of property together with an inadequate
increase in the rate of supply of that type of property, with the
result that the scarcity of such property increases. This is likely to
occur when there are population increases or increases in a
particular type of activity which increases the demand for specific
property types, such as houses or ultra-modern office
accommodation.

However should the opposite occur and the population
decreases or demand for a type of property decreases for other
reasons, the capital values of properties may decrease, particularly
if changes in value are measured in real terms rather than absolute
terms. There is nothing in the economic world which guarantees
that individual properties increase in value with the passage of
time, and it is quite possible that a trend towards more office work
being done at home could in time lead to a reduction in the real
value of central business district office accommodation, while rapid
development of computer shopping might result in falls in the
values of many retail property investments.

An investor should therefore only assume that a reverse yield
gap implicit in the price paid for a particular property investment
will be covered by capital growth if all the circumstances at the
time the investment is made indicate that there is likely to be real
capital growth attached to that particular investment. Emphasis
has been put upon growth in real terms as being the appropriate
measure as, in the 1970s and 1980s in particular, the capital
appreciation which appeared to accrue to many property
investments was in fact often nothing more than inflationary
growth. Many investors who were able to make substantial
financial profits from such growth will not have been concerned by
the fact that there may not have been growth in real terms, the

capital profit realized having been all-important to them. However, inflation is now seen as a major evil by most governments with government policies in many countries being based on a desire for either no or only low inflation, and the likelihood of capital gains being made as a result of high inflation will be reduced in the future if government policies aimed at restricting inflationary growth are successful.

Property investors and society

Overall it should be appreciated that property investors perform a useful role in society by assisting the property development process through, in many cases, guaranteeing the existence of an end-purchaser for a development which enables it to proceed because of the reduced risk carried by the developer. They also provide a pool of appropriate properties which are available for society in general to use on the basis of periodic rental payments, so releasing capital for employment in the business activities of many property users. In addition they enable the use of a wide range of properties by many individuals and organizations with limited capital who would be denied access to suitable properties if the only means of access was through outright purchase. In these respects property investors are far from the parasites they are sometimes painted to be by some. Indeed, they form an important part of the property development sector of the economy, and thereby fulfil an important economic and social activity for society in general.

Where property investors are publicly owned companies they also assist small investors (who might otherwise be unable to purchase individual property investments because of their high cost), to invest in soundly based property investments, so assisting saving and sound investment on a wide scale throughout society. When property investors and investing organizations also combine property ownership with a high level of stewardship exhibited through high quality management and maintenance, and appropriate upgrading and redevelopment of property as and when needed, they do indeed fulfil a most important role for society in general.

Chapter 5

The Development Team

While the prime motivator behind a development scheme will be the developer, many modern developments are so complex that it would often be impossible for a developer operating alone to complete them. Even if completion of some proved possible by a "one man band", it is likely that there would be many inefficiencies which resulted from the failure to employ, as and when appropriate, those with specialist skills. Accordingly, most projects in the modern world will be undertaken with a developer as the driving force employing a development team which may well include twenty or more different specialist consultants in order to achieve maximum efficiency and to maximize profits.

It is not uncommon with large projects for a development to be so complex that twenty or so different consultants could be employed merely to draw up a planning application. However, in the interests of the financial efficiency of a project, it should be the objective of a developer to restrict the number of consultants employed to those who are absolutely essential for the project to achieve the maximum success possible and to be financially efficient. So in a scheme which merely involves the erection of a small number of houses, "the development team" may in fact be restricted to the developer and an architect, with one of them also acting as project manager. The general approach to a development team should perhaps be that no consultant should be employed unless they can be seen to "add value" to the project in that without their use the scheme is likely either not to proceed or to produce a lower net profit. Consultants should consequently be employed as and when needed and when the profit produced from using them exceeds the cost of hiring their services.

The *developer* is obviously the most important member of the team being responsible for the initial concept and for transforming it from a concept to a completed development. He or she will ultimately determine, albeit with the benefit of advice from others, the type of scheme to be undertaken, the size of the scheme in both physical and financial terms, the amount of money which will be borrowed to fund the scheme, the location of the development and the site that will be purchased, the consultants to be employed to assist in creating the development, and all the other major decisions which have to be made with respect to the development.

There is no standard development team, and for each development there is likely to be an ideal team which will vary from scheme to scheme depending upon the size and type of scheme and a whole range of other considerations such as site conditions and planning conditions, and as indicated above the developer will ultimately decide the composition of each specific development team. At various times, depending upon circumstances, the specialists considered below are likely to be called upon by developers.

Property valuers will regularly be important members of a development team as they should have the skills and knowledge required to enable reliable estimates to be made of the likely market value of the completed development, and to do initial and advanced appraisals of the project. Their knowledge of market rental values, current market yields, and capital values is also important for these purposes. They can, in addition, advise on the value of potential development sites both in the open market and for the purposes of the proposed development, for these two values will not necessarily be the same. They can advise on the current use values of potential development sites to identify the minimum bid figure which current owners are likely to consider accepting, and they should have reliable knowledge of construction costs, the cost of employing other consultants, and the cost of finance, all of these being important inputs to the development appraisal.

Their knowledge of user needs and the critical elements of effective demand, coupled with their knowledge of investor requirements and investor interest and activity, should enable them to be key advisers to the developer on many aspects of design, and on matters affecting market value and the final marketing of a development.

Well-informed valuers should be in close contact with those in the real estate market-place and aware of developments in that market which might affect the financial viability of a project, and they should in particular be aware of other development proposals which are likely to be in competition with a proposed project. As well as being able to advise on virtually all factors which may affect the value of a proposed development, they will be able to undertake all the financial appraisals required by a developer, as a result of which they will often undertake the role of project manager, (discussed further below), financial control being an important task for the project manager.

Planning consultants are able to advise developers on the type of planning application that should be submitted and which is likely to find favour with the local planning authority as they are in regular contact with local authority planning officials and aware of their objectives and the constraints within which they have to operate. They will be important members of virtually any development team, but they are likely to be particularly important with larger schemes when negotiation with local planners regarding a whole range of considerations relating to design detail (and especially with respect to onerous conditions) may be an important part of the project design and may have considerable implications for the final value of the completed development.

Their ability to negotiate on planning conditions in order to contain the cost of compliance for the developer, and their ability to negotiate for sufficient development to maximize the value of the final development are both important, whilst their detailed knowledge of planning law and their understanding of planning objectives and planning systems equip them to deal with such matters better than most other consultants.

With anything other than very small schemes, the *architect* is likely to be an extremely important member of the team, and even with small schemes the architect may still be an essential consultant. As well as often undertaking the tasks described above under the role of the planning consultant, architects will be responsible for the detailed design of a scheme to ensure that it complies with all the planning conditions, that it complies with building regulations, that it satisfies the brief supplied by the developer, and that it will satisfy all the objectives of the development proposal, including the need for it to be built within specified cost limits.

The architect will receive a "design brief" from the developer which instructs him or her on the type of development to be designed, the style of development required, the amount of accommodation to be constructed, and the required quality of the building. This will include the quality of the exterior construction and finishes, the internal accommodation and all fixtures and fittings, the quality of such things as central heating and air-conditioning systems, lifts and escalators generally being very clearly specified. The developer should also specify the cost limits within which the architect has to design the development.

It should be the overall objective of the architect to design in order to ensure development within the cost limits at the same time as complying with the planning and building approvals, while the design should also provide a building which is efficient in use to maximize market value. If a design should result in the cost limits being exceeded or the building being inefficient in use, it may well result in the development not being a financial success. The architect's role is therefore crucial for the financial success of a development project. The architect may well provide the basic plans necessary for a planning application, and will provide the detailed plans necessary for building regulation approval and for the use of the building contractor in constructing the approved development.

Architects often act as project managers for schemes, but their education and professional operations may not necessarily equip them particularly well for the difficult tasks of financial control and personnel management, which are likely to be major tasks for project managers on big schemes, whilst it may be difficult for them to wear two hats, that of the architect for the scheme and that of the project manager to whom an architect would normally report. However, there are undoubtedly architects who are competent in these areas and also competent project managers.

Quantity surveyors are employed to provide detailed estimates of the cost of actually constructing a development. They work from the plans drawn up by architects and they cost the various elements of labour and material required for the development, including the cost of sub-contractors and items specified as "prime-cost items", to provide "bills of quantities". These are used to make an estimate of the total cost of constructing the buildings, together with the costs of ancillary work, such as landscaping and fencing, and amounts for contingencies which may arise during the construction period. Their objective in undertaking this task is to provide the developer with as reliable an estimate as possible of the likely total cost of completing the construction work.

As well as doing this the quantity surveyor can advise on the type of materials to be used for various elements of a building, and on the equipment to be installed in a development, such as lifts (elevators) and escalators. They will regularly advise the developer on the most suitable type of construction contract for a scheme, and will draw up appropriate contracts, after which they will invite building contractors to tender for a scheme, this usually being done on a selective basis taking into account the requirements of the scheme and the contract, and the capabilities of the various contractors. Following this they are likely to be important advisers in the actual selection of the building contractor by the developer.

During the construction process the quantity surveyor will cost work completed and work in progress, for in anything other than very small schemes the building contract will provide for the contractor to be paid portions of the total contract sum as various stages of the construction work are completed. Having calculated the value of the work done by the contractor and having verified that it is done to the appropriate standard, the quantity surveyor will advise the developer on the amount of the appropriate stage-payment to be made to the contractor.

In undertaking these various tasks the quantity surveyor will work in close liaison with the architect, particularly when decisions have to be made that work done is not up to standard and has to be re-done, stage-payments for work done generally either being withheld or reduced in such circumstances.

The quantity surveyor is another specialist who is also frequently employed as a project manager.

A wide variety of *engineering consultants* may be engaged by developers depending upon the type and size of development, and such considerations as the physical features of the site and the locality. Specialist engineers are most likely to be employed when development on difficult sites is involved, when developments are large-scale, when the building design creates difficult design and construction problems, or when the development incorporates the installation of specialist equipment.

Site engineers may be utilised to report on the quality of a site for construction purposes, its load bearing characteristics (particularly when these are not consistent throughout a site), the site drainage characteristics, the type of shoring necessary to adjoining land and buildings (if necessary) and for foundation construction, the depth needed for foundations, and the type of foundations which should be used.

Structural engineers may be used to advise on such matters as the appropriate type of framework for large buildings, the appropriate size of structural members (the size and type of support columns and beams in particular), and the design of floors to carry specific loads.

Heating and ventilation consultants will be required in particular when buildings are large and need such installations, when buildings are likely to be used by varying numbers of people (in which case systems will need to be designed to cope with variations in demands on those systems), and when buildings are partly open to the elements on either a regular or occasional basis. The task of such consultants will be to ensure that there is an adequate supply of fresh or treated air to the accommodation, that the air is changed sufficiently often, that the temperature of the air is appropriate to the needs, and that overall the quality of air in the buildings is such that those using the buildings can do so in an efficient and effective manner.

The problems for such consultants are not necessarily easily solved, and difficulties are created by the sheer size of many modern buildings, the large amount of glass used in many constructions, the variations in temperatures caused by such things as draughts, the varying numbers of people occupying rooms causing variations in total body heat in them, variations in temperature caused by the use of lighting and machinery, and high levels of heat gain and loss in some local climates. In seeking to design suitable systems engineers will have to take account of the developer's requirements and cost limits, statutory and public health requirements, the capital cost of equipment, and the cost in use of equipment, including the cost of cleaning and maintenance. The solution of these problems is an important stage in the design of a development as, in climates which are subject to extremes of weather in particular, the failure to install suitable heating and ventilation equipment may result in the value of a property being adversely affected to a substantial degree.

Lift (elevator) and hydraulic engineers play an ever increasing role in the design of buildings because of the sheer height of many modern buildings. Whilst views from the 25th floor of a high-rise office building may result in high rental values, this will only be so if sufficient high-speed lifts are installed to make it easy and quick for staff and visitors to reach those floors. Such consultants therefore play an important part in the design of high-rise buildings in particular, in that they are required to provide an adequate number of sufficiently fast and sufficiently large lifts to ensure that each floor is readily accessible, but they have to do so without creating unneccessarily large cost burdens for the developer. They also have to ensure that the service needs as well as the staff and visitor needs are provided for.

Overall, whilst operating within cost limits, lift and hydraulic engineers will have to ensure the provision of sufficient appropriately sized lifts in suitable locations, with appropriate load bearing capacities travelling at suitable speeds and built to an acceptable level of quality. In doing this they will have to do a careful balancing act between user needs, user desires, capital costs, running costs, reliability, and overall quality, whilst safety aspects have to be considered at all times. An additional consideration is that every additional lift tower which is designed into a building not only increases the cost of construction, but it also reduces the amount of space available for letting. This may in turn reduce the value of the building unless the increased capital value which results from the installation exceeds the cost of the lift.

The skills of these consultants are also likely to be required in the design of shopping centres when the transport of shoppers, employees and goods between different levels is important. In such developments there is likely to be a need for escalators, personnel lifts, and service and goods lifts.

The heavy dependence of modern buildings upon electrical installations and equipment makes the employment of an *electrical engineer* in a consultancy capacity virtually essential. Even homes now make substantial demands upon electrical installations, and the design of circuits and fittings which are appropriate for user needs will be even more important in the large-scale office, retail and industrial developments which are now commonplace.

The task of the consultant will be to design appropriate electric installations to cater for all user needs with appropriate access points for those needs. This will have to be done within cost limits, and there will be a need for close liaison with specialists such as the heating and ventilation consultants and the installers of other equipment which will itself place heavy demands upon the electrical installation. The design of a system which is adequate for likely peak usage is desirable, and at all times the requirements of safety laws will have to be observed.

The use of *catering consultants* may be necessary in such developments as office blocks, education complexes, entertainment, sports and recreation centres, and hotels to ensure that the provision of facilities for the large-scale catering requirements of such buildings is adequate, appropriately located, and built within cost limits. These consultants will need to balance capital costs with costs-in-use whilst staying within the cost limits, and they will need to liaise closely with other consultants such as the architect and the electrical engineer.

Information technology consultants are particularly important in the design of office accommodation, but the spread of information technology into other activities such as retailing is occurring at a rapid pace, and their expertise is probably required for any medium to large-scale development, so dependent has the modern world become on modern systems of information transfer. Their task is likely to be to ensure that the provision of circuitry is adequate for all potential user needs, with sufficient and appropriately located access points. They will need to work closely with the architect in particular to ensure their advice is incorporated into the design.

Interior design and furnishing consultants are likely to be used to advise on such things as the "user-friendliness" of finishes and colours, the quality, functionality, durability and costs of finishes and furnishings, and on health and safety aspects of the various finishes and furnishings available in the market-place. Their skills are becoming more generally appreciated with the growing realization that certain colour schemes may assist worker productivity with others being counter-productive, while some finishes and furnishings carry health and safety risks because of such considerations as the risk of fire and the emission of noxious or unpleasant fumes.

They may also be important advisers with respect to the marketing of property: their ability to decorate and furnish office and residential accommodation in a way which persuades would-be users to pay high rents or high capital values for accommodation can be a marketing tool which helps to enhance returns to a development.

Landscape architects may achieve the same results as interior designers for a developer through the design of lawns, gardens, flower-beds and shrubberies and other landscape features in a way which considerably enhances a development and the returns produced by it. Property in a pleasant environment is always likely to command higher values than similar property in a less pleasant environment, to the extent that nowadays, even with industrial and warehouse developments, good landscape design can considerably enhance the market value or the marketability of a development.

Marketing consultants may be used to ensure that a development is marketed in a manner which is likely to maximize the returns to it and to minimize the period of time required to successfully dispose of it through letting or sale, or both. With small-scale developments it may be that a real estate agent performs this task, while in the past decade or so the larger firms of real estate agents have recognized the importance of appropriate marketing techniques to the extent that some employ their own marketing specialists competent to handle large-scale developments.

These consultants should ideally be used from the design stage of a development as their advice on design features which are likely to assist marketing may be invaluable. They would also be able to start the marketing programme from the design stage; when "pre-letting" or "pre-sale" of space is required before a project is likely to be begun, their early use may be particularly advisable. Overall they will aim to ensure that marketing occurs at the right time, using the most appropriate marketing methods, and in a cost-effective but successful way such that immediately the development is completed it will be occupied and producing maximum returns for the developer.

With any other than small developments a *financial consultant* is likely to be a most important adviser, such a consultant dealing with a range of issues such as tax considerations and the overall financial control of a project. It may be that the developer or a project manager undertakes such tasks in smaller schemes, but, whoever is responsible, it will be necessary to ensure that all actions taken are tax-effective, that sufficient finance is available as and when required, that finance is raised as cheaply as possible, that loan terms are not unduly onerous, that finance is used in a cost-effective way and loan costs are minimized, and that there is an acceptable relationship between equity and debt finance. These matters will be considered further in Chapter 12.

The benefits which result from sound financing arrangements and from sound financial control cannot be over-emphasized, and there must be many examples of marginal schemes which have been very successful as a result of sound financial control, and likewise many otherwise good schemes may have failed through a lack of good financial control.

It is difficult to imagine a development project which will not require the assistance of *legal advisers*, and even the most humble scheme is likely to call for the services of a *solicitor* to undertake the

development project, and the ultimate profitability of a project is very much in their hands. They will therefore also need to have a thorough understanding of the financial aspects of projects and must be able to retain tight financial control of a project at all times.

To summarize the duties of the project manager, he or she is ultimately responsible for ensuring that the project runs without any hitches and that the development delivered to the developer is exactly as planned and contracted for.

The above list of specialists is not exhaustive, and a developer may be well advised to recruit the services of other consultants as and when necessary. Clearly, from the developer's point of view, the recruitment of specialist consultants is a considerable cost item (the total costs for consultants on some schemes being very large indeed) and, as indicated earlier, they should only be used when they either result in reduced costs or increased returns which more than compensate for the cost of hiring them. When using consultants the developer may be tempted to recruit those who charge the lowest fee rates, but such a policy may be false economy as the ability to charge a high fee-rate regularly indicates that a consultant is extremely competent. The use of highly competent consultants, even at high fee-rates, may prove to be an economy in the long run, as opposed to using cheaper consultants who have lower levels of competence.

The need to have competent and reliable consultancy skills and to ensure that they are available as and when required results in many development organizations actually employing the more commonly used consultants on a permanent basis. Some of the advantages in doing this are that, as long as there is sufficient work to keep them fully employed, it will usually be cheaper than recruiting such skills from external consultants, whilst they will be available to be part of the design and development teams on a permanent basis from the day a development possibility is first considered. They will also become familiar with the needs and aims of the organization and will be accustomed to working with other permanent members of the development team.

The use of appropriate and competent consultants is likely to make a big impact on the financial success, or otherwise, of a project, and, in particular, the recruitment of a suitable project manager is likely to be a major step in ensuring the best possible results are achieved from a development project.

Chapter 6

Planning and Property Development

The fact that public sector planners have an important part to play in the property development process is generally accepted in developed societies, but the questions exactly what their role should be and how much power planners should have give rise to a wide range of differing answers. Indeed, if one considers the planning systems and powers provided by legislation in different countries, one sees different approaches to these two particular aspects of planning control. So, for example, there has for some considerable time been far more attention paid in the United Kingdom to detailed aspects of development control (although in recent years there has been some relaxation of this aspect of planning), than in Australia where in New South Wales, for instance, there has been a system of control in force which has placed less emphasis on development control at a detailed design level. Even though the New South Wales system is based in many respects on the British system, it has in general allowed individual developers much more freedom with respect to the detailed design aspects of development than has the British system, notwithstanding the fact that both systems have similar broad objectives.

Despite the above observations there are still some who maintain that there is no need for a system to control property development, arguing that market forces would ensure that the right amounts of the right types of property would be supplied through the free market, and that the same market forces would ensure that the design and quality of construction of developments would be of a standard acceptable to society. It is further argued that a supplier of property would have to pay regard to market needs to be financially successful, and that no supplier of such an expensive product could in any event afford to take undue risks by supplying something that was not clearly demanded by those able to pay for the product in the market place. The large financial commitments made by developers would create a situation in the market in which they would do their utmost to ensure that the product was exactly what the market wanted, otherwise they would face financial disaster.

Additionally it is argued by some that planners employed in the public sector are, in any event, not necessarily able to judge what is best for society as they are removed from the market-place and are therefore not conversant with public demand. They are also influenced by a range of ideologies and pre-determined objectives of planning control which result in them not being unprejudiced judges of what is best in development terms, while they are also likely to make decisions based on their own subjective opinions which are not necessarily in the best interests of society in general. Indeed, their views may frequently be strongly influenced by political motives rather than true planning objectives. Property developers, on the other hand, are in regular touch with the market-place and well aware of the needs of society as expressed through those demanding property. They cannot allow their own subjective opinions to dictate their actions, as to do so could cause financial failure for them.

Many of these arguments can be put into context by considering what happened in most countries before development controls were introduced. More recently the incredible problems experienced in the property development industries in a range of countries in the early 1990s cast doubt upon the ability of many developers to accuratedly read the market-place. Certainly, it would be difficult to argue that all developers in the second half of the 1980s acted on sound economic principles or in ways which were in the best interests of society in general.

The rapid development of industrial towns and cities in the United Kingdom during the Industrial Revolution occurred in the absence of a modern system of planning and resulted in the indiscriminate location of developments, residential and industrial properties being developed alongside each other in a way which inevitably resulted in there being an unhealthy environment for the residential occupiers. The problems at that time were particularly bad because most of the industrial operations gave rise to the emission of smoke and fumes in contrast to many of today's "high-tech" industries which are not so socially objectionable.

Development during the Industrial Revolution also tended to occur irrespective of whether there was an adequate infrastructure, whilst in any case the capability to supply modern sewerage and water supply systems often did not exist or was not considered important by developers. Many developers were in fact industrialists, more eager to provide some form of living accommodation to house workers at low cost to themselves (to avoid reducing their eventual profits) than to provide high quality residential accommodation for their fellow members of society. There were, however, notable exceptions to this such as the Cadbury family, who developed the model village of Bournville on the outskirts of Birmingham, England, and others. The main objective of many Industrial Revolution developers was in fact to assist the maximization of profits from their own activities as industrialists, and the finer aspects of property development of the time were probably unknown to them and of little concern to them.

Property development in the free market of that time resulted in sprawling and rapidly growing towns and cities, developed with little or no concern for such things as well-planned road networks, the appropriate location of conflicting uses (eg residential, social and recreational, and industrial activities), the provision of well-constructed and well-designed buildings, and the provision of adequate services to properties. To argue that the same would not happen all over again in a free market situation would be to ignore the fact that exactly the same has occurred in many of the recently developed urban areas of Third World countries in Africa, Asia and South America, and also in some of the countries of Eastern Europe. In these countries the desire for growth and the absence of adequate development control has resulted in urban sprawl, the development of shanty towns, and the growth of urban areas without the adequate provision of services and communications systems. As a result, economic inefficiencies and health and social problems exist, this despite the fact that mankind now is aware of the dangers posed by such problems and despite the fact that it has the technical knowledge and ability to avoid repeating many of the mistakes of the past.

The ability of property developers to supply just what is needed, in the right locations, and of an acceptable quality is also cast into doubt by market conditions in most property markets of the developed world in the early 1990s. Cities such as London, Sydney, Melbourne and most of the major cities in the USA and Canada suffered from very large over-supplies of central business district office accommodation, with vacancies in such accommodation in various cities being anything from 10% to 40% of the total available supply. Such over-supply was not restricted to office accommodation and in some areas there was an excess of hotel, retail and industrial accommodation also.

Clearly developers completely misread the market and devoted scarce resources to expensive developments for which there was inadequate demand. There are many reasons why this happened, and amongst them is the fact that many developers probably pay inadequate attention to macro-economic factors, being more influenced by short-term and local demand factors than by long-term, national and international economic considerations. The international downturn in most economies around about 1990 therefore caught many of them by surprise and left them with developments on their hands which turned out to be unwanted in the market and therefore financial disasters. This problem was almost certainly worse than it might have been had property developers even taken note of what their immediate competitors in the market were doing, but it is suspected that even when they are aware of competition most developers operate in the belief that their own developments will be so good that they will inevitably be successful. To operate with such an attitude is to ignore the fact that value only results from utility, that relative scarcity affects the level of value of any product, that the level of demand affects the level of value, and that the level of supply does the same.

The question of timing is important and was perhaps not considered enough by developers, because no matter how good a particular development may be, if it reaches the market after most demand has already been satisfied it is likely to have only limited success, while if it reaches the market during a recession, even the most superb property may be a financial disaster. It is easy to be wise in retrospect, but it appears clear that many developers failed to consider "the bigger picture" as a result of which soundly designed and well-constructed developments actually came to the market at a time when there was no effective demand for them.

Such basic economic considerations are fundamental to the success of any product in the market-place and yet warning indicators which existed in the 1980s appear to have been consistently ignored by developers. High and increasing levels of unemployment and problems of controlling inflation over a long period of time were evident in many countries, some of which also suffered from little or no economic growth over a number of years. Additionally, in many countries high or frequently fluctuating interest rate levels and unstable foreign exchange rates added further uncertainty. Despite this uncertain economic backcloth, the total amount of planned new property was in many cities very substantial indeed. It is small wonder that, with the advent of the recession or depression, in many cities there was subsequently much vacant accommodation; many new developments coming onto the market with little or no hope of being anything more than partially occupied at best; many partially-developed properties where construction activity was suspended; and many "holes in the ground" where, fortunately for the developers, expenditure did not progress beyond site preparation.

Such happenings caused financial problems for developers but they also caused considerable problems for society in general. Strains were put on the banking system as a result of defaults on loans for development and property investment. Property development overall suffered from a stigma which is likely to result in many desirable and necessary developments being viewed with suspicion in the future by financiers and the general population. The "knock-on" effect of property failures adversely affected many other areas of economic activity and, on the urban scene, there were eyesores which resulted from the existence of unoccupied buildings, partly completed buildings, and the "holes-in-the-ground" which could remain as blights on urban landscapes for many years.

For the above reasons alone it is believed that the case for a planning and development control system is well justified. Public sector planning departments, being composed of humans, will not get everything right and will inevitably make some mistakes, whilst the influence of politicians on the system can result in the imposition of policies which are not always in the best interests of the wider populace. However, their existence should ensure that the considerations of society are taken into account in most cases and in most aspects of the property development process, although under many existing systems it would not necessarily be possible for them to ensure the avoidance of many of the economic misjudgements made by developers in recent years. In some planning codes planners are in fact prevented from judging development proposals on economic and commercial grounds, their areas of control being restricted to the areas in which they specialise such as the consideration of overall urban design features, infrastructure provision, communications provision and design, the location of specific land uses, urban design on the micro scale, and the social aspects of development.

It is arguable that in view of:

(i)   the experiences of the past and the very substantial problems that have resulted from the over-development of certain types of property;

(ii)   the very large development schemes which are proposed nowadays;

(iii)   the large amounts of money involved in such schemes;

(iv)   the very long time periods which such schemes inevitably entail;

(v)   the large strain that big developments place on existing infrastructures and the high cost of increasing infrastructure provision to cater for them; and

(vi)   the rapid changes which can in fact occur in the overall supply situation if new and competing developments are allowed to proceed without consideration being given to their effect on the existing commercial scenario

the role of planning and development control should in fact include a searching consideration of the financial, commercial and economic implications of development proposals. To extend the role of planning and development control in many countries to include what would be this new area of activity would necessitate a considerable change in existing philosophies and approaches, but the problems experienced in many countries put a new perspective on the concept that the commercial implications of development proposals are not valid planning considerations.

To take on board such a role would, in many cases, require the acquisition of a new range of skills and the acceptance of different attitudes by many existing planners and planning departments. The importance of the profit motive in the property development process in particular would have to be accepted if sensible and socially desirable development schemes were actually to proceed to completion, this aspect of development being considered later.

It is considered that the achievements of planning systems since they were first introduced into countries such as the United Kingdom and Australia are sufficient to prove the justification for planning and development control, even though as with any discipline or system they have their deficiencies. Apart from the viewpoints forwarded above, to have no system of planning control, leaving development decisions to be influenced by market forces, might well result in developments proceeding which were against the interest of the majority of society, there being no guarantee that those able to pay for a product are representative of society in general or sensitive to its needs.

It is now appropriate to consider in slightly more detail what benefits are to be obtained from land use planning and what the objectives of planning systems should be with respect to property development.

The objectives and roles of planning and development control

A well-designed and well-administered planning and development control system can ensure that:

(i)   development does not occur in inappropriate locations and that inappropriate types of development do not occur;

(ii)   the type of development required by society is allowed for in the planning process thus facilitating its provision;

(iii)   liaison and co-operation between different planning authorities takes place, for example between neighbouring local planning authorities, and between local and regional authorities;

(iv)   there is a reasonable balance between "over-planning" which is too restrictive and "under-planning" which fails to adequately achieve the overall objectives of planning control;

(v)   there is a sufficiently democratic system to allow suitably qualified, well-informed, and genuinely interested people to have inputs into the planning system;

(vi)   there is an acceptable quality of planning and design in development which is actually carried out;

(vii)   there are acceptable spatial relationships between different types of land uses;

(viii) there is adequate control of infrastructures to ensure that public services are suitable and available for development when required;

(ix)   suitable density controls are enforced to prevent over-development and the over-intensification of uses in different areas of development;

(x)   there is suitable design control at the local level to exert a positive influence over individual developments, thus improving the general standards of individual developments and neighbourhoods;

(xi)   there is suitable quality control at the local level to ensure that development which does take place is of an acceptable quality of construction.

It is possible to assign other objectives to a planning and development control system including the achievement of a range of social objectives, but for a consideration of the property development process it is suggested the above objectives are the most important, and each will be further considered in turn.

The prevention of development in inappropriate locations and the prevention of inappropriate development in any situation are important functions of a planning and development control system.

Locations which would generally be considered as inappropriate in the normal course of events would include:

(a)   those which would result in there being an unnecessary extension of development into currently undeveloped and unspoilt areas;

(b)   locations which ought properly to be protected as areas of great natural beauty and locations in which development would be harmful to areas, sites, or buildings of significant social, historical, cultural or architectural interest;

(c)   locations in which suitable infrastructure provision does not exist or in which an existing infrastructure would be over-stressed by further development, the upgrading or extension of the existing infrastructure only being possible at unreasonable expense;

(d)   locations in which existing uses would be adversely affected in an unreasonable way by further development;

(e)   locations in which development would prejudice the further reasonable development of an area in which further development would in fact be acceptable at a later date – for example, development which if allowed now would at a later date restrict access to otherwise usable back-land;

(f)   locations in which the granting of development permission in respect of an individual development proposal, which in itself would be reasonable would, nevertheless, represent an undesirable precedent that could stimulate other applications in an area in which development was in fact generally undesirable.

Examples of inappropriate development might include:

(1)   development which would be incompatible with adjacent activities in terms of the type of use to which it would be or might be put, particularly if such use would be likely to pose a threat to health or safety;

(2)   development which is likely to be incompatible with neighbouring developments in terms of acceptability of design;

(3)   development which would result in the over-intensification of the use of a particular site to the detriment of existing land uses, or in a way which would place excessive strain on the local infrastructure or essential services;

(4)   development which would be likely to generate an unreasonable amount of traffic resulting in increased noise and fumes, or strains on the local road system;

(5)   development for which land is allocated in other more suitable locations;

(6)   development for uses for which there is already sufficient and suitable accommodation available in appropriate locations.

Making provision to ensure that the type of development required by society can in fact occur is one of the most important roles a planning and development control system can fulfil.

Unfortunately, in reality there is too often a negative approach to planning rather than attention being paid to the positive role it should play for society in the property development arena. An effective system should in fact:

(i)   allocate sufficient land for development to enable the land use needs and development needs of society to be satisfied;

(ii)   allocate such land in appropriate locations for each specific type of land use in terms of compatability with, co-operation with and the avoidance of conflict with activities carried on in other land areas and in other buildings;

(iii)   ensure that appropriate land is allocated for various uses sufficiently far in advance of the need for new buildings to enable them to be constructed in time to satisfy user needs;

(iv)   ensure that appropriate infrastructures of roads and other essential services are provided to land reserved for development sufficiently far in advance of the development of individual sites to enable development to occur in time to satisfy user demand.

Planners should in fact play a very positive role in the property development process, and many of their objectives should be very similar to those of property developers. Planners and developers ought in theory to work together in a very co-operative way in pursuit of similar ends, but in reality they too often work in adversarial ways. Both planners and developers should be seeking to ensure that the types of development required by society to enable it to function in a way that the majority of society considers appropriate, can in fact be built by developers in suitable locations for prospective users, to a standard of design and construction and provided with all the amenities and facilities that are required by would-be users, hopefully at costs which enable users to purchase and use the finished product.

Too often the role of planners is commonly regarded as being negative and preventive whereas the major task of planners should in fact be positive: that of ensuring that acceptable development can occur at an appropriate time, in suitable locations to assist the maximization of productivity and the provision of satisfactory working and living conditions for society in general.

Liaison and co-operation between different planning authorities is extremely important and desirable if an overall system of control is to function efficiently and to be respected by those affected by it.

Sadly this does not always occur, and there have been instances of adjoining planning authorities operating policies which cause conflict or a lack of consistency at places where their boundaries abut each other. Similarly, instances have occurred of authorities taking over planning responsibilities from other authorities with immediate changes of policy which adversely affected many whose own earlier decisions had been based on the policies previously in place.

Clearly changes in policy are likely to occur when there are changes in those administering systems, but there is a definite need for such changes to be carefully thought out and justified before they are implemented. Changes which are perceived as being implemented simply in pursuit of sectional interests or which are thought to be purely politically inspired are unlikely to be acceptable to many in society. For any system to be respected and its provisions to be observed by the general populace it needs first and foremost to be seen as being in the interests of society in general and as being equitable.

No system is perfect and it is inevitable that there will be instances in which there are inadequate liaison and co-operation between different control authorities, or changes in policy. Where this happens it is likely that individuals or organisations may be adversely affected, and in such circumstances it is desirable for there to be an appeals procedure in place, over and above the appeals system necessary for the normal operation of a planning control system. Such a system should adjudicate on instances of alleged injustice and, where appropriate, there should be powers to make exceptions to policy and, as necessary, to provide for compensation where hardship can be shown to have occurred. Without such provisions there are likely to be instances which cause public disquiet to the extent that an entire system of control may be thrown into disrepute, even though in other respects it may be acceptable.

To minimize problems of this nature it is important that there should be good "horizontal co-operation", that is good liaison between neighbouring authorities to ensure that there is consistency and compatability between their policies at least where their boundaries abut each other. There is likewise a need for good "vertical co-operation", or good liaison between each control authority and the authority either above or below it, again to ensure that their policies are consistent and compatible. Apart from the avoidance of conflicting policies, good liaison between authorities ought, in theory at least, to result in the avoidance of over-provision of facilities, particularly those which are publicly funded. Such things as sports facilities and infrastructure facilities can be used jointly by those in neighbouring authorities, and it is logical in view of the cost involved in providing such facilities that they should be put to maximum use and that they should only be provided where there is a need.

The situation, that existed in parts of New South Wales in 1995 with respect to retail property development, provides an example of the lack of co-operation between neighbouring authorities which in the long run is likely to benefit no-one. At a time when a major objective of many local authorities was to attract new sources of employment in the hope of reducing unemployment and increasing economic activity in their areas, planning approvals were given by different authorities for the development of new shopping centres to the extent that there was likely to be a considerable over-supply of new shopping centres on a regional

This will assist in avoiding the impression that a system is imposed on society from above by a remote and uncaring administration. Many amongst the population in general are well-informed, particularly when local matters are involved, and are well able to make valuable contributions to the consideration of development proposals. Whilst it is generally to be hoped that professional specialists are able to give the best and well-balanced advice on many matters, they do not always see things from a wide enough perspective and the views of the well-informed lay-person or local resident can be useful inputs into the decision making process. Planning should be beneficial to society in general and should not be imposed for the benefit of a government or any particular interest group. For a system to be seen to be democratic and equitable there must be adequate opportunity for those whose properties are affected by development proposals to make representations to those in charge of control. In recent years it has generally been accepted with many control systems that the types of individuals and groups who should have the right to make representations should be considerably extended.

Developments may have effects over a far wider geographical area than the immediate vicinity in which the development is proposed, and to restrict the right to make representations to immediate neighbours only may be unreasonably restrictive and may prevent legitimate representations from being made. People in general are now very much aware of factors which affect the environment and matters such as ecological considerations are now often raised when development activity is proposed. Such matters may be very important, but it is likely that they can only be adequately expressed when specialists are given the opportunity to be heard. Similarly, where buildings or localities of special historical or architectural interest are likely to be affected by development proposals, it is again likely that the full implications of the proposals can only be adequately considered if special interest groups or individuals are allowed to express their concerns about the proposals.

The problem with democracy is that it can be very inefficient and very time-consuming in its application and administration, and the danger with extending the right to make representations in respect of development proposals to too many people is that the development process becomes too prolonged with unacceptable increases in costs. It is for this reason that it is believed that, although there is a wide need for the power to make representations, such power has to be sensibly controlled both through the initial composition of rules and regulations, and through their sensible and efficient application. The concept of a completely democratic development control system in which all have the right to participate has its attractions, but in reality it is likely to be so costly and time-consuming that it would be impracticable to implement such a system.

Apart from the time and money considerations of such a system, there is the danger that too much democratization merely encourages the development of pressure groups which have their own sectional interests at heart rather than the interests of society in general. Although, in theory, such a system would provide more opportunity for local residents to be heard in respect of specific development proposals, this is only likely to be of benefit if those residents are articulate or are represented by professional advisers, who may be too costly for local residents' groups to afford. In the event, it is suspected that where more democratic provisions allow greater representation, specific, limited-interest pressure groups possibly benefit at the expense of local residents whose views and needs often do not get adequate representation or adequate consideration.

The development of more democratic planning control processes and their administration is also a costly exercise for the public sector and will inevitably result in a larger administrative body. Even though many development control systems now attempt to recoup costs through charges for development applications, it is likely that, unless charges are to become prohibitively high, only partial recoupment of costs will be possible. If this is so, greater democratization of the system will result in increased burdens on public purses, which generally are very limited. It will also result in increased costs to applicants caused through delay and the need to respond to what probably would be a larger number of representations on applicants' proposals.

Despite the above facts, it is critical that a planning and development control system should give adequate opportunity for public input, particularly at the formative stage of policy and plans, if it is to be widely accepted by society for whose good it is intended. Many lay people may have different viewpoints to professionals and specialists, and those viewpoints may be important and relevant considerations in the development control process.

A good control system should ensure that there is an acceptable quality of planning and design in development which is carried out.

A major objective of a planning and development control system should be to ensure that members of society can carry out their various activities in well designed, functional buildings in the most suitable locations and in an appropriate environment for each specific activity. Planning and design criteria and their application should seek to ensure that each activity of mankind can be carried out with the least possible damage to the natural environment and in a way that ensures the least possible conflict between the various land uses and various land users. They should also seek to ensure that a local environment and the developments therein both provide as pleasant and inoffensive an environment as possible in the interests of those living and working in it, and that individual developments are so designed as to be pleasing in appearance whilst functioning efficiently for the purpose or purposes for which they are intended.

What exactly constitutes an acceptable quality of planning and design is difficult to define as so many subjective considerations are involved, but a control system should attempt to ensure that in reaching the inevitable compromises which will be necessary to define standards acceptable to society in general, there should be sufficient scope to allow for variations in design. This will provide variety and the opportunity for those with design flair to be creative, whilst at the same time avoiding excesses which are likely to be found offensive by a significant proportion of the population. To achieve general planning and design criteria which are acceptable by all is probably impossible, and it could well be that the best measure of success or otherwise is how few or how many of the population find the standards to be unacceptable. In that respect many modern planning codes could probably be adjudged very successful, as while there are likely to be many individuals who complain about particular circumstances which they believe have adversely affected them, in general terms the majority of the population in countries with modern development control systems would probably agree that the overall benefits produced by the systems greatly exceed the problems caused by their existence.

The protection of the environment has assumed a particularly important place in public policy in many countries in the latter part of the twentieth century as research and improved knowledge have created a greater awareness of the damage done by humans to the environment over the centuries, but in particular since the start of the Industrial Revolution in Europe. Not only has the prevention of further damage to the environment been given emphasis, but considerable attention is also being paid in some countries to the improvement and re-use of land which has been damaged in the past by development, or which has become abandoned and disused as a result of urban blight. The cleaning up of land which has been contaminated by such activities as the storage or use of chemicals, or by the carrying out of certain industrial processes thereon, and the retrieval of land previously used for quarrying and mining are now major objectives in many countries. Quite demanding requirements with respect to such operations, may now be attached to planning approvals. Where this is the case, there may be considerable cost implications for those contemplating development activity.

Such measures have been spurred on by the realization that land is a resource which will not produce maximum economic benefits unless it is properly used, while in specific locations and in specific categories of land type it may well be that there is a scarcity of supply relative to demand. The realization that the mis-use of land and developments on it can also result in major health and safety threats has created a situation in which efforts are being made to remedy many of the mistakes of the past and to avoid repeating them in current and future development.

A planning and development control system should seek to ensure that there are acceptable spatial relationships between different types of land uses and land users.

This objective is generally addressed by drawing up planning policies and development plans or land use zoning maps which designate different areas of land for different types of development or land uses. As already observed, during the Industrial Revolution in the United Kingdom it was not uncommon for houses, heavy industries, shops and properties designed for social and cultural activities (for example churches and schools), to be developed "cheek by jowl" in a way which resulted in unhealthy and unpleasant environments. Usually none of the land uses was provided for adequately, and the efficiency of each use was impaired by the existence of incompatible adjacent uses and conflicting activities carried on in neighbouring developments.

It should be a major objective of modern planning systems to minimize conflict between different land uses and land users, and early planning schemes allocated different areas of land to different land uses. Development plans had "residential zones", "industrial zones", "commercial zones", and so on, providing specific areas of land reserved for development for each specialist land use to ensure that the sort of conflict that once existed between uses, such as residential and industrial uses, did not occur in the modern, planned environment. The general objective was to segregate conflicting land uses and, judged against the historical backcloth, there was sense in the objective.

However, some of the most pleasant residential areas are to be found in villages and small country towns, where historically there has been a random juxtaposition of land uses, a situation which probably could not have arisen under most modern planning and development control systems. The major difference between the environments in such settlements and the larger towns and cities was probably that the former did not have the heavy and dirty industries which the latter had; the intermixing of different land uses therefore did not cause the same level of danger, bad health, and generally unpleasant living conditions as existed in the larger, industrial settlements. The difference in the total scale of development was probably also significant, in that what may prove acceptable on a small scale in a small town may well be completely unacceptable on a larger scale in a larger town.

However, early planning systems tended, not unreasonably, to concentrate on the segregation of different land uses, and this placed a heavy emphasis upon the provision of adequate routes and means of communication between the different use zones. Residents had to travel distances to work which were often quite considerable being in a different land use zone, and they had to do so in many cases either by public transport or by private car, the distances frequently being too great to be covered on foot or by bicycle. Similar problems existed with going to the shops or taking children to school, although in most development plans attempts were made to ensure that primary and junior schools were located within easy walking distance of all parts of residential neighbourhoods, only older children needing to travel to school by vehicle. One of the results of zoning policies as practised in the past has therefore been to very much increase the dependence of most citizens upon travel by road, either by public transport or by car.

Such policies have therefore been traffic generators, which have contributed to what is perhaps the biggest problem faced by many modern planners – controlling and providing for the private motor car, and controlling the pollution produced by motor vehicles.

It is ironic that in many areas the older, objectionable and unhealthy heavy industries have now disappeared having been replaced by modern, high technology industries which produce little noise and very few fumes and smells. It is arguable that many of these modern industries could be suitably located either in or on the fringe of residential areas, so avoiding the excessive use of motor vehicles with the dangers they create and their great use of scarce natural resources, and also reducing the levels of pollution caused by their use.

Despite these observations and reservations, a good planning system should ensure that:

(i)   there is a proper relationship between conflicting land uses to ensure that mutually incompatible uses are segregated from each other to the benefit of the users, society in general, and the environment;

(ii)   traffic generating uses are located close to means of communication and with easy access to major roads in particular, to minimize the dangers they create and the inconvenience and harm to other land uses;

(iii)   noisy and dirty uses and those which give rise to noxious smells and fumes are segregated from other land uses, and from residential areas in particular;

(iv)   uses which may be a danger to health and safety are located well away from other uses, and that any dangers created by them are thereby minimized;

(v)   where development is allowed to occur, the density of development is controlled to ensure that land is not over-developed to the detriment of the local environment in general;

(vi)   land is allocated for the types of use which might otherwise be ignored if land uses were determined by market forces alone (for example for hospitals, schools, churches and charitable uses).

The allocation of land for uses which might be completely ignored if all land use decisions resulted from bidding in an open market situation is an important consideration, as some uses which have high social values would almost certainly be neglected if such a situation ruled. Society would be the loser if this were the case, and a good planning and development control system can help to ensure that minority uses and socially desirable uses can be provided for, irrespective of financial considerations. Whether land allocated for such uses is actually developed and used for those purposes may well be dictated by market forces, but a planning system can at least ensure that land is in fact available for such uses and that sufficient land is designated as open space, so helping to improve the environment in urban areas. If public funds are made available for the purpose it can also ensure that much of the open space is dedicated for use by the general public as "public open space".

One aspect of the spatial control of development is the protection of the countryside and undeveloped areas in general. The British system of planning places a high priority on this aspect of planning, and the creation of "green belts" around urban areas was from the early days of modern planning an important feature of the system. In green belts there is a general presumption that only in very exceptional circumstances will any form of development be allowed other than limited types of development for such purposes as forestry and agriculture, and the result of the creation of green belts has been that the encroachment of urban development into rural areas was controlled. Not all systems have placed the same emphasis upon the protection of the countryside, and in New South Wales in Australia there only appears to be limited control of development in the countryside. There is, as a result, much urban sprawl which encroaches into the countryside and there is almost unbroken urban development along the main routes out of Sydney for as much as forty or fifty miles.

Such would seem to be an undesirable form of development leading, as it almost inevitably does in due course, to the over-stressing of what are essentially rural infrastructures on the ever-growing fringes of the city. The demand for a more appropriate infrastructure then arises, but such infrastructure provision may be dictated by the forces of previous, relatively unplanned development, rather than being provided in a form which would have been more suitable and more efficient and effective if the previous random style development had not occurred.

It is interesting to note that in the United Kingdom in 1995 the use of the concept of green belts was reaffirmed as a major tool in

Town and Country Planning, such reaffirmation resulting from the widespread acknowledgement that the concept has indeed resulted in major overall benefits with respect to the preservation of the countryside.

A planning and development control system should seek to ensure that there is adequate control of infrastructures to ensure that public services are suitable and available for development when required.

Property development can only take place in an efficient manner if there is an adequate infrastructure and if all the services required by modern society are available. It should therefore be a function of a good planning and development control system to ensure that the provision of both the necessary infrastructure and public services is planned in advance of need, and that they are actually provided in time to enable development to take place as and when society requires an increased supply of properties.

The provision of road networks and, if appropriate, other transport facilities such as railway lines (which could be particularly relevant in mining and industrial areas and with large-scale new residential development); sewer services; water, gas and electric supplies; and telephone services is essential for modern living and working requirements to be satisfied. Their provision takes time and is costly, and therefore requires adequate and sufficiently detailed advanced planning, together with budgeting to ensure that the necessary finance is available when needed. Large-scale development activity may require major infrastructure provision such as the construction of dams and reservoirs, or major land drainage schemes, before any development at all can occur.

The provision of an appropriate infrastructure and public utilities requires large sums of capital and it is consequently most important that once provided they are used in a way which maximizes their efficiency and effectiveness. It is also best that they should be provided in areas or in ways which will result in the greatest possible returns to the capital required for their provision. Such factors may be major considerations in determining which areas are most appropriate for development and the programming of areas for development. It may be that there are areas which in other respects seem suitable for development, but the high costs and difficulty of providing an infrastructure mean that in overall financial terms they are really unsuitable, although with the passage of time and changes in circumstances that situation might alter.

Not only are infrastructure services and public utilities costly to provide, they are invariably also costly to run and maintain and this is another reason for ensuring that there will be planned maximum use of them within a reasonably short time frame once they are provided.

It should consequently be an objective of planning to seek to avoid the provision of services for only a few likely users, and development in remote or isolated areas or to a limited scale should be resisted if it is likely to result in requests for the unreasonably expensive extension of existing services or the provision of new services at unjustifiable expense. There may of course be situations in which this cannot be avoided, but at the very least it is arguable that in such circumstances the users who benefit from the provision of costly services should in fact cover all the costs involved in their provision.

In summary, it should be a function of a good planning and development control system to ensure that desirable development is assisted by the provision of an appropriate infrastructure and appropriate public utilities sufficiently far in advance of development need to ensure that the properties required by society can be provided by property developers at the right time, developed to appropriate standards, and supplied at acceptable prices.

A good planning and development control system will seek to control the density at which development is allowed to take place to ensure that there is not over-development which may both place unacceptable strains on an infrastructure and public utilities, and result in unacceptable living or working conditions in a locality.

Clearly, if development of too intensive a nature places strains on an existing infrastructure or public utilities in a locality, both existing users and the newly developed properties are likely to suffer from an inferior level of service, which would be highly undesirable. There seems to be an inborn wish on the part of many property developers to put as much building on any particular site as is possible, supposedly in the belief that such a course of action will maximize returns to the developer. However, this is not necessarily the case as building values may be adversely affected if there is not an acceptable level of services for would-be users. During the 1980s in Dublin, Ireland, there were new office blocks which were unlettable as the Irish telephone service was unable to supply telephone lines to them, so rendering them unusable for business purposes. The redevelopment which had occurred had been in excess of the ability of the telephone authorities to service the area, and it could well have been the case that the construction of smaller or fewer properties, which the telephone authorities were able to service, might in fact have created higher net values for developers in view of their subsequent losses and costs incurred by them through holding completed but unlettable properties.

Development or redevelopment which is so intensive that it provides more space than the amount which can be adequately serviced by local utilities is likely to have a limited appeal to would-be users. All other things being equal, it is certainly likely to be less attractive to potential occupiers than property that has a higher standard of services. The capacities of local mains supplies and their ability to cope with proposed new developments should therefore be an important factor in the consideration of planning proposals by control authorities; if excessive development is permitted both existing and new users are likely to suffer from inferior service provision. Any development which is allowed should therefore be within the limits with which existing infrastructures can cope, unless upgrading of services is possible, and density control measures can be implemented to ensure that this is the case.

Over-intensive development can also have the undesirable result of reducing the amount of daylight and sunlight which can enter buildings and city streets, so reducing the qualities of both internal and external environments. Controlling the amount of building which can be placed on any one site can help to ensure that there are adequate daylight and sunshine levels in urban areas, particularly when there is also the ability to control the siting of individual developments so permitting the shadows cast by each building also to be taken into account.

Delays caused by traffic congestion waste a considerable amount of time and are therefore costly, and the over-intensive development of an area to the extent that vehicles cannot move freely or cannot be easily parked when necessary is undesirable. Where two or three storey buildings are demolished and replaced by new properties many times their size, over-intensive development of an area can occur with resultant operational inefficiencies. Many more vehicles may come into the area than was previously the case, and unless there have been road improvements it is likely that roads that were suitable for the previous developments will be quite inadequate to cope with the newly increased traffic loads. The time lost through traffic congestion can be very considerable and may result in loss of productivity and lower returns to organisations than those that could be achieved in the absence of congestion. It could again be the case that less intensive redevelopment of a site or an area could in fact reduce or eliminate such problems, and that the total values of properties might in fact be higher (or at least the total profits from redevelopment) because of the resultant higher quality of environment, this in turn producing higher rental and capital values per unit of accommodation.

Whilst density control measures can help to control the quality of an urban environment, when properly applied they are likely to help to maintain acceptably high efficiency levels and so to enhance property values at the same time. In that respect development at densities which are too high should be something which both planners and building owners should be anxious to avoid.

A planning and development control system should seek to ensure that there is suitable design control at the local level to exert a positive influence over individual developments, thus improving the general standards of individual developments which, all being well, will in turn assist the general improvement of neighbourhoods also.

It is suggested that good design control should, more specifically:

(i)   ensure the overall attractiveness of individual buildings and ensure that any adverse impact of new development on adjacent properties or localities is minimized;

(ii)   attempt to ensure that new development is compatible in design with adjacent properties, or at the very least that it is not incompatible to an unacceptable extent;

(iii) result in the choice of suitable and compatible building materials;

(iv)   seek to ensure that any new development is of an appropriate scale to blend in with existing development;

(v)   ensure that there is overall harmony in the appearance of an area; and

(vi)   ensure that new development takes place in compliance with desirable health and safety standards.

It is probably much easier to state objectives such as these than it is to define precisely their meaning at operational level, or, furthermore, to define how they should in fact be achieved. While most people would say that good design is important in any development, different individuals will have different concepts of what is meant by the expression. One only has to consider the fact that some purchasers will pay huge sums of money to buy fashion clothes which others would never dream of purchasing or wearing, to appreciate that design control at local level involves decisions being made on a wide range of matters on which it is almost impossible to avoid subjective judgements being made on matters over which subjective opinions are likely to vary greatly. Just as taste in clothes varies, so one person may suggest bright-blue roof tiles add desirable colour to an urban scene, whilst another might find them gaudy and objectionable. Any code providing implementation standards and rules on such matters is bound to give rise to debate, but even if there is not complete agreement in the general populace as to what is desirable or acceptable with respect to local level design control, it is believed there would be few who would seriously argue for the complete abolition of design control criteria and regulation in the planning process, to be replaced by a state of affairs in which any developer could design exactly what they wished in terms of the appearance of a building and the materials from which it was constructed.

From a developer's point of view, even if specific design control rules introduce constraints they will, nevertheless, have some benefits in that they will define those constraints and indicate to a developer at an early stage of design what is likely to be unacceptable and what is likely to be acceptable. By doing that, design rules or guidelines should reduce the amount of abortive design work with which developers might get involved, so controlling costs.

It might be argued by some that improving design costs money, and that the existence of design control rules therefore increases the cost of development. It is not axiomatic that good design costs more than poor design, although better designers are likely to charge higher fees than poorer designers. However, good design can result in lower costs in the use of property, lower maintenance costs, and a more efficient design for the ultimate user which results in greater operational efficiency. It may therefore be the case that even when good design results in increased initial capital expenditure, that may be more than compensated for by savings in use over a period of time and by increased capital values.

A planning and development control system should seek to ensure that development which does take place is of an acceptable quality of construction. (This objective will be considered separately in the following chapter.)

It is appropriate to comment on "planning gain" and "development contributions". It is indisputable that, where there is development control, the granting of planning approval creates a situation from which developers can make profits, whilst if no approval to develop is granted, such profits cannot be made. Many therefore argue that the ability to make those profits having been granted by society through the giving of development rights, part of those profits should be contributed by developers to society through what is known as "planning gain". So it has frequently been made a condition of development approvals that developers should undertake works for the public good which were not originally part of their schemes. Approval to develop a shopping centre might carry with it the condition that a building has to be provided for use as a public library or that a certain amount of public car parking space has to be provided. Indeed, some planning codes make legal provision for such development contributions, and in New South Wales the legislation provides for what are known as Section 92 payments by developers which might be used, for instance, to improve roads in the locality of a new scheme, to provide car parking facilities, or to improve local surface drainage.

Developers might argue that unreasonable demands are made by many planning authorities, but there is little doubt than in the past many developments have occurred which have placed costs on society in general which ultimately have had to be met by local authorities, such costs resulting directly from the impact of new development but not having been borne by the developer. There is a delicate balance in which, in equity, developers should meet all the costs which result directly from their schemes, whilst on the other hand, if unreasonable and excessive financial demands are placed on them by planning authorities, development costs may increase so much that a particular project becomes unprofitable and will not proceed.

Any additional cost which increases a developer's total costs of development will generally reduce the potential profit from a development and, if cost increases result from the demands for contributions by planning authorities, the latter should be fully aware of the financial implications of their demands to the financial viability of a project. If demands are excessive they may result in schemes being abandoned, and if the project was for a scheme which would have benefited society in general, then society may, at the end of the day, be worse off than if lesser demands, or no demands, had been made for development contributions. Whilst most would agree that it is only right to require developers to meet all public costs that result directly from their project, requesting a "donation" from the anticipated profits raises other considerations, and may be a positive deterrent to development activity in areas in which such requests are regularly made, particularly when they are excessive. Developers may well decide to divert their activities to localities in which they regard attitudes and demands as being more encouraging for developers, to the overall loss of the area in which excessive demands were made.

It should not be forgotten when considering this delicate topic that, if developments are financially successful, developers should in any event pay part of their profits to society through the normal tax channels. The fact that some may be able to minimize tax payments through these channels probably results from faults in tax systems or tax collection systems, and the correction of such shortcomings should, in the writer's view, be the responsibility of politicians and tax departments rather than of the planning system.

General observations on development control

Overall, while there may be extensive debate about the precise objectives of development control regulations, desirable rules and regulations, and the most effective and acceptable ways of implementing a control system, few would disagree that the threats posed by uncontrolled development are so great that controls on development are indeed essential in the modern world. The degree of control which is desirable is likely to give rise to debate, as are the objectives of controls, and it should be stressed that the suggestions in this chapter are the views of the author. Others might disagree with some of these views and might suggest a range of other objectives and other desirable forms of control, and it would be extremely difficult to select even a small group of people among whom one would be likely to get complete agreement on such matters.

However, planning control is here to stay in most countries, although the precise criteria and rules may vary slightly from time to time. As indicated earlier, without development approval, development cannot legally occur, and liaison with public sector planners, understanding the constraints within which they work and the objectives they seek, and reaching agreement with them in both the informal stages of project planning and the formal development application stage, are important tasks for any developer. Such tasks require tact, understanding, and the ability to design within publicly prescribed limits to achieve a financially acceptable development.

Chapter 7

Building Control and Property Development

As suggested in the previous chapter, it is desirable to ensure that any development which does occur is to an acceptable quality of construction, and in most countries it will be necessary to obtain a building or construction control approval, in addition to a development approval, before development can legally be begun. Development consent usually is given if an application satisfies a broad range of planning and development considerations such as those discussed in the previous chapter, while building approval generally depends on matters which are more related to the actual form of construction and the construction process. Matters such as the site and foundation works proposed, the form of construction proposed, the detailed design features of the building, the construction details, and the materials to be used in construction will be closely considered at this stage to ensure that they comply with the relevant building code for the area in which it is proposed to locate a development.

Building codes and regulations vary widely from country to country for a variety of reasons. Depending upon a range of considerations such as the topography of an area, soil types and sub-soil conditions, the climate and seasonal variations in weather, the possibility that an area may be subject to such things as high winds, heavy seasonal rainfalls, heavy snowfalls, long periods of excessively hot weather, drought, or earthquakes, so will the design requirements for a particular locality be drawn up by those responsible for determining required standards of construction and ensuring compliance with those standards. There may, therefore, be a need for considerable variations in building regulations within relatively short geographical distances, whilst in large countries there are likely to be different climatic regions which result in standards appropriate for one region being quite inappropriate for another.

Another important factor likely to determine the level to which building controls are developed is the state of development of the economy of a country. It is generally true that the existence of building controls will result in the costs of development being higher than they would be in the absence of those controls for at least some of the development carried out in a country. The general objective of controls being to ensure that minimum standards of construction are achieved, development which would otherwise have fallen below those standards will either be improved to comply, or will not proceed because of the increased costs which compliance would invariably entail. Those responsible for determining the standards to be imposed should therefore recognize that the imposition of controls, or the introduction of stricter controls, is likely to increase the costs of development and consequently to result in the abandonment of some development which would otherwise have occurred. That is an acceptable result if such development not occurring is considered the lesser of two evils, as opposed to it proceeding but to an unacceptable standard of construction.

It should not be forgotten that there may also be wider economic considerations in that a reduction in the total amount of property constructed may result in shortages of supply of certain types of property in the market-place. Where property is nevertheless constructed but at higher cost, there may be price implications in the market place if the supplier is able to pass on those higher prices to those renting or buying property. There may be further economic implications if the latter group is in a position to pass on price increases to others, the result possibly being increases in the prices of a relatively wide range of goods and services.

Clearly, the more affluent a society and the higher the general level of income within it, the more that society is likely to be able to afford high standards of building construction, and vice versa. However, even in relatively affluent societies, it should not be forgotten that one reason for the existence of homeless people or those who live in houses which are considered to be unfit, is that often they cannot afford the minimum price for housing which the standards of building control indirectly determine. There is therefore the "trade-off" that ensuring desirable construction standards are met may result in some people being "priced out" of the market. What may be an admirable desire to ensure that all people have good quality housing may, in effect result in some having no housing at all, unless other steps are taken to improve their purchasing position in the housing market.

The general objectives of building controls are usually to ensure that:

(i) building standards are such that minimum acceptable health standards are likely to result;

(ii) minimum standards of construction are observed so that buildings are sufficiently durable and do not deteriorate too rapidly;

(iii) buildings are in general respects safe places to live and work in and offer reasonable protection from fire in particular; and

(iv) buildings are so constructed that collectively they provide an acceptable environment with a reasonable relationship between different building units.

Much of the early legislation with respect to building control resulted from the need for governments to try to ensure good standards of health among their citizens. The Industrial Revolution in the United Kingdom in particular resulted in many who had previously lived and worked in rural areas moving to live and work in urban areas where new industries sprang up rapidly. To cater for these new workers, industrialists often built housing close to their factories and mines. Such housing frequently was built very rapidly, to very poor standards of construction, and with little regard for health and hygiene, there often being no proper sewers to service houses and very limited supplies of water for them. In addition, the density of development was regularly extremely high with houses being built "back to back" with at best only small areas of garden, sometimes none, and little or no public open space.

Public health problems resulted which in due course led to the enactment of a series of statutes intended to improve conditions in general, the major act of the nineteenth century in the United Kingdom being the Public Health Act 1875, which gave local authorities the responsibility for public health administration in their areas. Housing legislation was also enacted, and since the early legislation there has been regular updating of requirements in respect of housing and building in general, such that there are now clearly stated and precise controls which are usually rigidly enforced to ensure that good standards of health and safety are achieved in all building work.

It is interesting to note that despite the passing of many years and the increased knowledge of people today, developments not dissimilar to those of the last century in the United Kingdom have occurred in Eastern European and Third World countries in the last quarter of a century, indicating the need for building control laws even in modern times.

The imposition of minimum acceptable building standards will almost inevitably increase the construction costs of some buildings above the cost of construction which would apply in their absence, otherwise there might be little point in having the controls. Controls which merely endorse what in any event is occurring are superfluous, and only those which increase the standard of construction really merit their existence and the existence of mechanisms to ensure they are complied with, for control systems are generally expensive if they are to be effective. Depending upon the level of development of a country and the severity of the controls in force, so compliance may be relatively easy in terms of cost for the majority of developments, or it may alternatively pose a major cost issue.

Once a building control system exists it is almost inevitable that the requirements of the system will be regularly and systematically increased to respond to instances of construction or building material failure that may occur, or health and safety problems which are revealed over time through the use of buildings. In the relatively recent past in various countries construction requirements have been increased to make buildings more able to resist earthquakes, whilst in many countries there has been regular updating of the regulations with respect to protection against the danger of fire in buildings. With respect to the latter, regulations frequently stipulate the number of external doors which must be provided in certain types of building; the number of staircases and fire-escapes to be provided, both internal and external; the types of fire-resistant building materials that may be used and the types of materials which create particular dangers in the event of fire and which must not be used in various situations; the need to install fire resistant doors in certain locations; the need for the installation of smoke detectors, sprinkler systems and other fire-fighting devices (particularly in high-rise buildings), and the need to install fire alarms.

The specifications for such works may be quite precise with inevitable cost consequences, and even though application for building control approval will normally be left until quite late in the development process because of the cost of preparation of appropriate plans, it is important that any potential developer should be aware of the general implications of building control regulations in the area in which they are working and for the type of development they propose. This knowledge at least forewarns them of the type of requirements with which they will have to comply and gives them an early indication of the likely cost implications. With respect to fire regulations in particular, early consultation with those responsible for enforcing the regulations is well advised.

With regard to the building control objectives suggested above, in the ideal world those objectives ought also to be the objectives of property developers. Even without statutory requirements and the need to satisfy specified design standards, in theory buildings built to satisfy those objectives ought to command higher market values. However, even should that be the case, it will not necessarily result in developers building to those standards of their own free will, as the developer's objective will generally be to achieve the highest possible profit, which does not necessarily result from the highest market price if costs necessary to achieve a higher price rise by more than the increases in market prices.

Whilst in theory, therefore, the developer should seek to build to acceptable building standards, in reality this may not happen without statutory controls for a variety of reasons which include:

(i)   purchasers often have limited perceptions of the features which make buildings acceptable and valuable, and are not necessarily conversant with appropriate health and safety requirements;

(ii)   despite the fact that purchasers of buildings often lack technical knowledge themselves and despite the high cost of property, they often fail to engage the services of knowledgeable specialist advisers, possibly because to do so adds extra cost to the inevitably expensive exercise of property acquisition;

(iii)   even when purchasers are well informed, the limited financial capabilities of many purchasers may result in them seeking cheaper properties despite the fact that they may be deficient in terms of quality of construction; and

(iv)   as indicated above, it may be possible for a developer to maximize profits by building to a low quality of construction – that is in the real market the profit motive may actually mitigate against high quality construction being provided.

With respect to the last factor, it may well be that a developer could make higher total profits by developing an estate of houses of poorer quality at low cost than by building better houses at higher cost (even if the quantity of houses were unchanged), since the cheaper houses would be within the purchasing power of a larger number of potential purchasers.

With respect to the objective of assessing the likely financial success of a possible development project, building regulation requirements can affect both the total amount of building it is possible to create on any specific site and the cost of creating that space. The total amount of development may be restricted by such things as the need to observe "daylighting requirements", the need to provide specified circulation space around a development, and the need to provide external fire-escapes on site, whilst costs can increase as a result of such things as the need to use specified materials and the need to provide a range of fire detection and escape devices in a building.

Building regulation requirements may also result in the total value of possible development on a site being restricted, in that the total size of development may be less than might otherwise be the case, while the amount of common space required in a building in the form of corridors, staircases, fire-escapes, lifts, and toilet areas may reduce the amount of lettable space which can be created. If both reduced final value and increased development costs result, the effect of building regulations will be to reduce potential profit, whilst if the costs in use of a building to would-be users are also increased, they will be likely to reduce their rental or capital bids for the property.

The combined effect of the various cost and value implications of building regulations can therefore be considerable, and developers need to be very conscious of the requirements and resultant costs when assessing the financial viability of possible projects. They also need to be very much aware of possible increases in requirements which may occur between the first concept and the likely completion of a project.

Chapter 8

The Availability of Land

Properties which are developed will only have value in the market-place if they are developed in locations which will satisfy the requirements of potential users. It follows that if developers are to be able to supply the right properties for users, land must be available for development in suitable locations. The provision of sufficient development land in suitable locations for the entire range of possible property developments is therefore an important part of the property development process. Without a sufficiently large supply the needs of would-be property users are unlikely to be satisfied. However, at this stage it must be stressed that the *suitability* of land for development purposes is even more important than the mere availability of land.

If a shortage of developed property exists, property developers and investors may not suffer too much as the result of a limited supply may be that prices in the market are enhanced, resulting in extra-large profits for developers and high returns for investors. However, society may suffer from a shortage of developed property as those who are fortunate enough to obtain properties may have to pay higher rents and purchase prices than would be the case if there was a larger supply. The overall economy may also suffer if property shortages deny would-be producers accommodation in which to run their businesses or result in the costs of production of goods being increased.

The role of planners in ensuring that there is a large enough supply of development land is therefore very important in both social and economic terms. Not only must land be available for development, it must be capable of development in physical terms, and it must be suitable for development in economic and legal terms, and in respect of a wide range of site requirements, otherwise developers are unlikely to be tempted to develop it. Suitability for development will be determined by a range of site characteristics.

Accessibility is very important and proximity to main-road networks, railways, local access roads and other forms of communication is likely to increase the development potential of a site. Such factors are likely to be particularly important for industrial, commercial and retail development, and the ease with which workers can reach industrial properties and with which raw materials and finished goods can be received and despatched will be important considerations. Likewise, good accessibility by public and private transport will increase the value of residential properties, while retail areas also need to be easily accessible by car and public transport.

The *location* of a potential development site will be a major factor in determining how suitable it is for a particular type of development. Location is a key factor in determining the value of completed developments, and different types of development have different locational needs. The ability of a site to satisfy specific needs will determine its value for development, and the potential profitability of a development will be greater the more suitable the location is for would-be users. It is therefore critical that land allocated for various uses is located in situations which are well suited for those uses in both physical and economic terms. If sites are economically unsuitable for a particular type of development, development is unlikely to prove profitable and in the normal course of events would not be expected to occur for that type of use. Those allocating or zoning land for different uses must therefore take into account the economic forces relevant to each particular use, and the site and locational needs of each type of land use.

Site access is a very important consideration; any development needs to have suitable approach roads and a suitable site entrance for the type of traffic likely to use the completed development. Access roads, which may be suitable for residential traffic, may be quite inadequate for an industrial area, and the same may well apply to site entrances. It should not be forgotten that construction traffic may be heavier than the end-user traffic, and if access is inadequate for construction traffic the costs of development are likely to be inflated.

The provision of sufficient and adequate accesses to a site is critical for the site to be efficient in use, and care should be taken to ensure that site access is sufficiently good to cope with peak usage traffic and pedestrian numbers as well as more normal working loads. If existing access is not adequate, the ability to acquire suitable access and the cost of acquisition are important in determining whether a site is really suitable for development.

The *size and shape* of a development site are important as size is likely to determine the amount of development which can be placed on site, whilst the shape of a site may affect the efficiency of the design of the building or buildings, which in turn is likely to have an impact on efficiency in use. Both may therefore affect the suitability of a site for development in economic terms in that they may well be critical determinants of the value of a completed development.

The *soil structure and load bearing capabilities* of a site may determine both the type of development and the quantity of building which can be placed on a piece of land. What may appear to be land suitable for intensive, high-rise development in view of location and other considerations, may in fact be suitable only for low-rise development when geological factors are taken into account. Whatever land may be zoned for, sub-soil tests may be a very important preliminary investigation to be made before a sensible development decision can be made for a site. Sub-soil tests should seek as carefully as possible to test all the area on which construction work is likely to be undertaken, as any unexpected excavation and foundation work may add considerably to the estimated costs of a development.

The *topography of land* may well be an important factor in determining its suitability for different types of development, and, for instance, all other things being equal, flat or gently sloping land is likely to be more suitable for development than land which has a steep slope. That is not to say the latter type of land cannot be developed, and in Hong Kong and other major cities such as San Francisco there are numerous examples of substantial developments which have been built on steeply sloping land. However, the economics of the market must have been such as to justify such development as the costs of construction would undoubtedly be far higher than those for similar buildings on flatter and more easily accessed sites.

Land drainage characteristics are important in determining the suitability of land for development, as, unless land has good natural drainage, a developer may be faced with the need to use an expensive type of design and construction and perhaps with having to install land drainage schemes, also at additional expense. Bad drainage characteristics may not always be readily apparent to the casual observer, and careful site investigation should be undertaken together with searching local enquiry.

The *availability of services* to sites or the ability to provide services within reasonable cost limits are important development considerations. If services such as mains water, mains sewers, mains electricity, telephone lines and also possibly gas supplies, are not easily accessible at acceptable costs, land is likely to have limited appeal to would-be users and therefore to developers as well. No matter how many other good features land may possess, without the availability of adequate services it will be unsuitable for development purposes.

The *legal interest available in the land* is of great importance in determining whether land is suitable for development. Generally developers will prefer situations in which the freehold interest in land can be purchased, and that such interests are unencumbered is likely to be another important preference. If only a leasehold interest is available, developers will generally wish to acquire a long leasehold interest, such a lease being sufficiently long to at least permit the full benefit to be obtained from all the development work which will subsequently be done on site. Circumstances and local practice will determine what is regarded as sufficiently long to make development worthwhile, and whereas in the United Kingdom developers like to get leases which are at least 99 years in length, and preferably longer, for example 125 years, in the Philippines leases considerably shorter than 50 years in length have been sufficient to encourage development in the expensive central business district of Manila. In the United Kingdom developers are seeking to achieve a situation in which, apart from the initial development operation, it may be possible to do a number of upgrades to the original development over the passage of time, followed by redevelopment of the site if the original development becomes economically or physically obsolescent. For such to be possible a minimum term in the region of 100 years is probably essential, whereas it would appear that in the Philippines developers have been prepared to develop on the basis of getting sufficient returns from the original development, plus one or two major upgrades, before the relatively short lease expires. In the latter situation the development valuation would need to be done in a way that ensured this was possible and that the ground rent agreed under the leasehold interest reflected the short lease term, so ensuring an adequate return to the developer or a subsequent investment purchaser.

Whether a freehold or leasehold interest in development land is available, a developer taking such land will wish to ensure that there are no encumbrances which either make it impossible to complete a sensible development or which result in any money returns from a development being unacceptably low. When restrictive covenants and third-party rights affect a site it will be necessary to ensure that they do not adversely affect the development or redevelopment potential, or alternatively that they can be purchased or removed at a time and a price which still enable a financially rewarding development to be undertaken. Where annual outgoings are payable on an interest in land, such as a rent for the interest or for rights over other land, it must be ensured that such payments are not excessive to the extent that they damage the development potential of a site.

Although it may appear that land is available for development, and although it may be suitable for development in planning and physical terms, if the legal basis on which it can be acquired by developers for development purposes is not acceptable to those developers in that a suitable profit is unlikely to be made from developing the land, then in economic terms the land is not suitable for development. It is therefore important that before land is allocated or zoned for specific types of development, consideration should be given to the existing ownership pattern of land and the types of interest under which it is owned. Such factors may be particularly important in determining whether it is suitable for one type of development or another, or alternatively in determining that, although development is in other respects desirable, it is impracticable because suitable legal interests cannot realistically be created.

The existence of an *approval to develop* is critical in countries or areas in which development control systems are in force. Without such an approval development cannot legally take place, and until such an approval is in existence it may be impossible for a potential developer to estimate accurately the amount of development or the type of development which will be permitted on an area of land. Both are critical determinants of development value, and the wise developer is unlikely to purchase land for development until the details of any development approval have been determined. In many countries land without a development approval will only be usable for its existing approved use, which may be very much less valuable than the value assuming it could be developed. Until a detailed development approval has been granted, land will remain unsuitable for development and the full implications of allocation in a development plan for a specific use must therefore be carefully investigated. It is essential to determine exactly what will be allowed and exactly how much space can be constructed on any site.

The precise type of development which will be possible on a site cannot be determined until a *building approval* has been granted, and this is likely to be a necessity in most countries. Details of the terms on which construction will be permitted will affect the type and amount of usable space which can be created, and the cost of creating it. As such, a building approval will be a determinant of the actual development value that applies (or does not apply if construction costs are likely to be excessively high), and until its details are known the precise suitability of land for development in economic terms will be unknown.

Apart from the above considerations, which are likely to influence a potential developer's opinion as to the suitability of land for development – a number of which affect the costs of development – these costs will also be affected by the amount of suitable land which is at any one time available in the market when one takes into account the fact that total development costs include construction costs, land cost, and finance costs. The more the amount of suitable land, the lower its market value is likely to be, and vice versa. Those responsible for determining the total amount of land available for development and the various uses for which such land can be developed, therefore have a major influence in determining the cost of development land, as if they restrict land supply they are likely to cause land values to be higher than if the supply was greater. This may not necessarily cause the value of developed property to increase, as the supply and demand factors for particular types of developed property will ultimately determine the values of different types of property, the cost of land merely being one factor in the supply side of the supply and demand equation. However, in some circumstances the high market cost of land caused by a shortage of supply may in fact result in substantial increases in the value of developed properties if, as a result of the constrained land supply and its high cost, sufficient new developments cannot be undertaken to cater for unsatisfied demand.

Development approvals

The fact that planning authorities can determine whether development will be permitted or prevented on specific areas of land will be a very important determinant of where development value will accrue on particular blocks of land, and on the amount of such development value where it does occur. The power to give or withhold permission to develop land, and also to determine the density at which development is permitted when it is allowed to occur, places planning authorities in a position in which their decisions determine the distribution of development value created by market demand.

The granting of development approval will not in itself make the development of land appropriate in commercial terms, and it will only be the existence of market demand for a particular type of development on suitable blocks of land which will result in the existence of development value in that land. For this reason the granting of development approval on land will not in itself ensure that the land is developed. There may also be instances in which development is deferred because a landowner or land purchaser decides to delay development in the hope of making a higher profit at a later date, or for other policy reasons. There will, in addition, be cases in which development is deferred, or perhaps never takes place, purely and simply because the land does not have physical characteristics which encourage its development, despite the existence of an approval to develop it.

Unless enough suitable land is allocated by planning authorities for development, sufficient development cannot take place which might have serious consequences. High market prices may result if demand for specific types of development is high but supply is restricted; this occurred in the United Kingdom in the second half of the 1960s when the government of the day restricted the development of office accommodation, particularly in Greater London. The government took the view that too much money was being devoted to office development which should, in its view, have been devoted to other forms of investment which would, it believed, have resulted in greater economic benefits for the country. However, the restrictions on new office development actually resulted in the eventual leasing of office accommodation which had previously been unlet, and in due course in higher rental levels as the previous excess of supply disappeared, while there was no real evidence of any diversion of investment into the activities considered more desirable by the government. The major result of the legislation, which was initially accompanied by rent restriction provisions, was in the first instance to cause crisis in the property investment industry which affected a wide range of people as it adversely affected the value of insurance policies and pensions. So great were the problems that the rent restrictions were in due course removed, and the major result of the legislation then turned out to be enhanced rental levels which benefited property owners and adversely affected property users.

Restrictions in supply brought about by a shortage of suitable development land may result in some people being unable to purchase or rent property because the resultant higher price levels actually price them out of the market. Such a state of affairs may have harmful social consequences if it results either in homelessness or in people only being able to afford sub-standard residential accommodation. There may also be harmful economic consequences if a shortage of accommodation and higher accommodation prices result in the restriction of desirable commercial and industrial activities in particular.

Another outcome of such a situation may be that those who are fortunate enough to own some of the land on which development permission is granted may make very high capital gains because of the enhanced value of developed property, and the resultant high value of development land which results from the restricted supply of developed property. The acquisition of high capital profits which result almost entirely from the fortuitous allocation of development approvals is considered by many to be undesirable and socially divisive, and may lead to accidental and large differentials in wealth amongst different members of society. Unless such a situation is accompanied by substantial taxes on such accidentally acquired profits it is difficult to justify in social terms.

The adequacy of the supply of land for development

It is therefore desirable for planning authorities to ensure that there is an adequate supply of land allocated for a range of specified uses to suit the needs of society in general. To do this public sector planners need to have a good understanding of the financial and physical processes of property development, and also the desirable spatial relationships of different types of land uses as determined by users as such considerations have important economic and social consequences. For example, a shortage of shops in a low-cost housing area because no land has been allocated for such a use, may result in those on low incomes having to pay high travel costs to shop, while the remoteness of residential areas from commercial and industrial areas may similarly result in high costs for those travelling to work.

It is important for planning authorities to try to determine:

(i)   the amount of land available at any point in time for specific types of development;

(ii)   the amount of land likely to be used in the current year (or other specified time period) for each type of development;

(iii)   the amount of land likely to be required in future years for each type of development; and

(iv)   the resultant total supply of land needed in the longer term for each type of development.

Having determined future land use needs, the authority is then in a position to allocate sufficient land for each purpose, to ensure its release with development approval at appropriate times, and to ensure that an appropriate infrastructure, particularly suitable access roads and services, is available at an acceptable cost in advance of demand.

A problem with such an objective is that the provision of services in an area too far in advance of demand results in expenditure on costly capital works from what are generally limited budgets before high use is likely to be made of those services. On the other hand, the provision of services such as mains water and mains sewer systems may only be financially justifiable if it is done on a reasonably large scale. To provide such facilities on an incremental basis as and when demand actually arises can in fact be an expensive method of provision, as there may be no economics of scale. The responsible authorities therefore have a difficult "balancing act" to perform which necessitates careful and regular monitoring of the adequacy of service provision.

The size of development blocks (or development plots)

The size of blocks (plots) of land allocated for each type of land use will need to be related to a number of factors, including the type of use and the size of the town or city if an urban land use is involved. So, by virtue of the type of use, industrial blocks will generally need to be large in comparison with residential blocks, while by virtue of the size of the urban area, a city centre office development block will need to be much bigger than an office development block in a small provincial town.

In all cases, however, blocks should be big enough to allow for an appropriate size of development in both physical and economic terms, and to provide adequate space for user access and service access (including the emergency services), while for some use types (particularly industrial users) it will be desirable for each block to be large enough to provide sufficient space for future expansion. However, caution has to be exercised against the extravagant allocation of development land, and, for instance, the failure to control the location of buildings on large residential blocks can prejudice the future intensification of land use as a result of the original building monopolising a block and also possibly restricting access to back-land.

The extensive use of land can also be expensive in that long lengths of road can result serving relatively few dwellings or other types of development, and long runs of services, such as water mains and sewers will also be necessary. Unduly long runs of this nature can result in unreasonably high infrastructure costs for each serviced development block, putting unreasonable and unnecessary pressure on public sector capital budgets, and probably ultimately resulting in higher development costs for each block developed.

Summary of considerations in the provision of suitable land for development

To summarise the above considerations, planning authorities in exercising their positive roles in the development process should seek to make sufficient suitable land available for each development need by ensuring that:

(i)   there is a sufficient supply of land allocated for each required development activity;

(ii)   there is an appropriate spatial relationship between the land areas allocated for different development purposes;

(iii)   land allocated for each use is in reality physically and economically suitable for that use;

(iv)   appropriate services are available at an acceptable cost for each development block in advance of actual need;

(v)   despite the above requirements, land is not used in an extravagant way; and

(vi)   similarly, the development infrastructure and essential services are also not provided in an extravagant way.

The use of virgin land for development

Developers frequently prefer to work on land which has not previously been developed, commonly referred to as "virgin land" (or "white land"), as it generally creates fewer problems than using land which has previously been developed. There are a number of reasons for this, including:

(a)   where a clear, undeveloped site is available development will generally be easier than if demolition and site clearance is required, and, accordingly, development costs should be lower;

(b)   there will be no delay because of problems caused by the existence of protected buildings (although environmental considerations might nevertheless cause delay);

(c)   virgin land is often in more attractive areas in physical terms, frequently being closer to attractive countryside and away from areas of urban blight and urban decay;

(d)   often, the use of virgin land for development will permit a less restricted design approach, for example road patterns are unlikely to be pre-determined and the need to adopt a design to be compatible with adjoining construction may not exist;

(e)   usually sewers and other services will be newly provided when virgin land is developed, and they are therefore likely to be more acceptable to both developers and property purchasers because they will be modern and built to the latest design specifications, mains should be of adequate capacity, and no renovation should be required for a substantial number of years; and

(f)   virgin land is also more likely to be located close to areas in which other facilities are relatively new and designed and built to the latest standards. Other properties such as neighbourhood shopping centres, health centres, and community centres may well have been provided recently to the latest design and health and safety standards, and to satisfy modern user needs.

There are, however, disadvantages in using virgin land for development which include the following:

(1)   Virgin land is frequently located on the edge of urban areas and its use on a regular basis can lead to an increase in urban sprawl, often at a rapid rate.

(2)   Over-large urban areas can result, with neighbouring areas often spreading to join each other to provide huge areas of continuous urban development.

(3)   Such urban sprawl will often result in the need for the construction of or extension of expensive communications systems, such as new roads or new rail links.

(4)   As urban areas increase in size the cost of provision of ancillary services may well become high because of the distances involved, for example the cost of postal services and refuse collection.

(5)   The concentration of development and the provision of services in new suburban developments may well lead to a shortage of finance for the maintenance and possible redevelopment of existing developed areas, particularly city and town centres, leading to the deterioration and eventual decay of those areas and their infrastructures. The development of suburban and out-of-town shopping areas, for instance, may result in a rapid loss of trade in established town centres, often resulting in empty shops which help to speed up the further deterioration of the affected centre. The migration of town centre residents to the more modern suburban developments exacerbates the problem as it results in a smaller residential population in the town centre to use the existing shops and other services.

(6)   Edge-of-town development frequently results in the loss of good agricultural land, although where modern agricultural techniques are practised this is not necessarily as great a problem for agricultural production as it was when less efficient farming methods were used. However, such things as established sports grounds and amenity land may be lost when development occurs, often to the disadvantage of a community.

The redevelopment of previously developed land

When the redevelopment of previously developed land is

contemplated, there will be a range of factors to be considered by the potential developer, some being of advantage but others being disadvantageous. Factors which may be considered advantageous include:

(i)   It may be possible to make use of existing road and service infrastructures, so reducing the cost of provision of public services which would benefit the local authority and service providers, the developer, property users, and property purchasers.

(ii)   Existing urban areas can be rejuvenated so countering any tendency towards urban decay.

(iii) The developer can develop in a known environment which may reduce the risks of development. There may be an established local market for the type of accommodation to be provided and, for example, modern office developments regularly occur by redevelopment in areas of established demand for office accommodation.

(iv)   From the perspective of potential property users, where redevelopment occurs in established areas they are able to determine, before committing themselves to take a property, such things as the general quality of the local environment and local services. This in effect helps to reduce the risk factor for property users in their selection of suitable properties, and accordingly should also help to enhance property prices for the developer.

There are, however, disadvantages from the developer's point-of-view in redeveloping in established areas, and to a large extent they mirror the advantages of developing on virgin land, including:

(a)   Existing street patterns may act as a fetter to design and may prevent, or make difficult and costly, the use of the most appropriate style or design of building to satisfy modern user needs. For example, existing building blocks as determined by the street layout may be too small to allow the development of appropriately sized buildings to adequately cater for total demand and also to benefit from economies of scale.

(b)   The problems and associated costs of unifying a number of different building blocks to create larger sites may be too great to justify a developer attempting such an exercise. (Some of the problems in site amalgamation will be considered further in the next chapter.)

(c)   The existing service infrastructure may be in need of extensive and costly renovation work or may even have inadequate capacity for modern needs and may therefore be in need of complete renewal with enlarged load capacity.

(d)   The cost of demolition of existing buildings and of site clearance may be substantial, and it may be difficult to accurately estimate the work which would be required in such an exercise. This may be particularly important if earlier site uses have resulted in site contamination which may be extremely expensive to remedy.

(e)   Where redevelopment would involve construction work on a cramped site it might be necessary for a developer to negotiate and pay for the rights to use adjoining land or the air space over such land during the construction period.

(f)   Existing developments and sites may be subject to restrictive covenants which may need to be removed to permit sensible redevelopment, and the removal or modification of such covenants could be both time-consuming and costly. It may also be that sites are subject to the rights of neighbouring landowners, such as rights of light and rights of way, which might restrict the type of redevelopment which was possible, sometimes to the extent that redevelopment would not in fact be financially viable.

(g)   It may be that the interests of existing lessees have to be bought in order to provide a freehold interest in possession to permit redevelopment to take place, this adding to the cost of obtaining vacant possession even if the purchase of such interests proves possible.

(h)   A combination of the above factors may make the redevelopment of inner city sites possible only at development costs which exceed the likely market value of completed developments.

The above factors may cause developers to seek permission to develop at high densities which may be well in excess of those considered appropriate by planning authorities and pressure groups in particular, and by society in general. The developer seeks higher density development in the hope that it will produce higher returns to cover increased costs, but the effect of more intensive development may not necessarily be good for a city or town in terms of urban design considerations and environmental considerations. In certain parts of London and in the city centre of

Dublin, strict rules have been implemented in the past in efforts to ensure that the urban environment is protected from the over-intensive development of city centre sites. For many people the very intensive development of inner city sites creates an environment which they find oppressive, it leads to increased pollution, and it also tends to result in buildings of great historic and architectural value being dwarfed by or lost among large, new buildings to the detriment of the townscape. There are a number of examples in Sydney, Australia, of historic buildings being lost among high-rise office developments to the extent that their true qualities are difficult to appreciate, while they are frequently not even noticed by passers-by, which in part defeats the reason for their having been retained in an essentially different environment.

Additionally, too much intensive development may place unreasonable strains on streets and services, with unacceptable levels of congestion at peak hours and overloaded commuter transport services.

Land holding public bodies and their effect on property development and property markets

The need to ensure that sufficient and appropriate land is made available for development is generally recognized by planning authorities and governments, although when the task is delegated to a large number of local authorities it may well happen that the collective product of those different authorities may be neither adequate or appropriate. There has been recognition of this problem at various times in a number of countries, and there have been attempts to organize the provision of development land at a regional or national level.

The Land Commission Act 1967 in the United Kingdom established a Land Commission ". . . to acquire, manage and dispose of land..." It was intended that the Commission would ensure that the right land was made available at the right time for the implementation of national, regional and local plans, and the Act also provided that a substantial part of the development value of land should accrue to the community through the collection of a "betterment levy". The objectives of making appropriate land available were sound, but in the event those objectives were never achieved for a variety of reasons. The legislation was complex, difficult to understand, and difficult to implement, and having been enacted by a Labour Government it was in due course repealed when a Conservative Government was returned to power.

Among the objectives of the New South Wales Department of Housing are those of increasing home ownership in NSW and increasing land availability. In the 1980s a land commission known as "Landcom" was established and it:

> ... develops, markets and sells affordable land, particularly in low-cost areas, to home buyers and builders. This involves retailing developed vacant land as homesites to individual purchasers, marketing land and house packages in conjunction with builders, financiers and service authorities and wholesaling underdeveloped land to private buyers, developers and builders.

It is probably true to say that the main objective in establishing Landcom was to ensure that land was acquired and partly developed by the installation of service roads and services to individual blocks in order to ensure that house blocks were available for would-be house owners to further develop by the construction of a house. In that respect Landcom was seeking to ensure that suitable land was available for development at the appropriate time, with a suitable range of services provided. It is interesting to note that it was set up by a NSW Labour Government, and the opposition Liberal Party, with policies which favoured more private sector activity, indicated that if returned to power it would abolish Landcom. In due course it was returned to power, and in the event the new Government decided against its abolition, with the result that Landcom continues in existence in the late 1990s and provides land for development onto the open market.

Whether provision is left to the free market or controlled by government, the availability of sufficient land for development is critical if adequate development is to take place, and if the cost of providing suitable developed properties for society in general is to be kept within the financial means of would-be users. Anything, such as a shortage of suitable land for development, which acts to restrict the supply of any type of property, is likely to result in the costs of property provision being unnecessarily high, which in turn is likely to have wider economic consequences.

Chapter 9

Site Assembly (or Site Amalgamation)

As implied in the previous chapter, existing sites or ownership blocks may be too small for redevelopment purposes to suit modern requirements in terms of the type of properties demanded in the market-place. It may therefore be necessary for a developer to amalgamate two or more sites to provide a site which is large enough for the development of a suitable modern building, and which also enables an economically sound development to be undertaken.The type of land block which was large enough in the past for office development may be far too small nowadays in view of the increased size of many organisations and also in view of modern administrative needs and techniques. Additionally, the piecemeal redevelopment of small city blocks may be physically difficult and costly, and hence unattractive to a developer in financial terms. To redevelop using existing building blocks would often result in the creation of modern buildings with small floor sizes and low efficiency ratios (that is there would be an unacceptably high percentage of total floor space devoted to uses such as stairs, lifts, landings, corridors and toilet blocks and other common space, which, although they are essential parts of a building, do not directly produce rental income). Additionally, there could be considerable difficulty on small development blocks in providing enough on-site car parking spaces to satisfy the parking requirements of both would-be occupiers and local authorities in many towns and cities.

The acquisition of a number of building blocks and their subsequent amalgamation to provide a larger development site could help to overcome many of these problems. However, there may be great difficulty in actually identifying enough available properties adjacent to each other to provide a suitable development site in a suitable location, while the cost of purchasing each site may result in total acquisition costs becoming high. In particular, this is likely to be the case if the individual owners of required property interests become aware of a developer's plans and needs, with the result that each may hold out for a high and often unreasonable sale price, that is unreasonable from the developer's point of view when the figures are incorporated in the overall financial appraisal of the envisaged project.

As a result, site assembly is only likely to be undertaken by a private developer if the possibility of a substantial profit exists, and even so the exercise must be undertaken with great care to ensure that potential profit is not dissipated in the site assembly exercise. One of the best examples of site assembly was recounted by Oliver Marriott in his book *The Property Boom* (Hamish Hamilton Limited, London, 1967, subsequently published by Pan Books Limited, London, 1969). He described the assembly of the site of the Euston Centre in Euston Road, London by Joe Levy, one of the best-known post Second World War London developers, and Robert Clark, a shrewd Scottish banker. Following an approach from a client who had a one acre site to sell to the north of Euston Road, Joe Levy devised a scheme for a possible thirteen-acre development scheme at the junction of Euston Road and Stanhope Street.

There were considerable risks as the proposal was to develop offices in a new area for office use, and site assembly would involve the acquisition of many interests. The area was described as shabby and:

> ... behind the frontage to Euston Road ... was a collection of decaying Georgian terraces and sad little shops. Joe Levy himself extravagantly described the area as "a derelict bloody den of disease".

The plan was to acquire the various property interests as quietly as possible and three agents were used for this purpose. The majority of the properties were bought in the period between 1956 and 1960, and the number of separate deals totalled 315. If any of the deals had fallen through Levy and Clark would have spent a great deal of money on land with no development potential and only negligible returns on the existing properties. There were large variations in the prices paid for the various interests, and some landowners had to be rehoused on other sites, these including a church, a hospital and a school.

In due course the entire site was available for the development of the Euston Centre which proved to be a very successful development. However, the development was only made possible by the courage, skill and patience with which the site was assembled.

If site assembly is undertaken it is wise to keep the development intentions as secret as possible, although this is likely to be difficult in an age in which planning approval has to be obtained and in which public participation in planning is commonplace. The advisability of entering into early informal discussion of proposals with planning departments, other government departments which may have an interest such as highways departments, and with other individuals or groups with a legitimate interest in specific development schemes, in the hope of getting both departmental and social support for a major scheme, tends to operate against the type of secrecy which is best observed in site assembly operations.

If development proposals cannot be kept reasonably secret a developer should seek to make arrangements with the owners of interests in land which are required to enable a scheme to proceed, but to defer all property acquisitions until a common date, so ensuring that no property is acquired until it is clear that the necessary approvals will be received, and also that all the essential interests in land can be acquired at acceptable prices. Such a result might be achievable through the use of options to purchase the interests which are required, or through the use of conditional contracts to purchase which would only become binding if all necessary interests could be acquired at acceptable prices and all necessary approvals would be granted.

Clearly, whatever devices are used to assemble development sites, the process is likely to involve long and often complex negotiations, and hard bargaining, and also to be costly and risky. Consequently, the process of site assembly is only likely to be undertaken if the potential profit is sufficiently large to justify the exercise.

Similar considerations may be relevant when only one block of land is involved but it is necessary to acquire several interests in the same land block before redevelopment (or development) is possible. It may be that there are, for example a freehold interest, a leasehold interest (possibly several leasehold interests if a property is in multiple occupation), and even sub-leasehold interests in the same block of land, all of which need to be acquired before a freehold in possession, or a sufficiently long leasehold interest to justify development, can be owned. In circumstances such as this in which the merging of legal interests is involved, it is particularly important that no owner should be left in the position of being able to demand an unreasonably high sum from a developer simply because the developer is already committed to purchase other interests. As with the acquisition of different blocks, the developer should use similar devices to ensure that no interest is acquired until the ability to purchase all necessary interests is assured.

The compulsory acquisition (or resumption) of land for development purposes

Public authorities are regularly given powers which enable them to compulsorily acquire land for purposes which are authorised by law; such purposes often include property development and redevelopment. Many authorities have the power to acquire land for their own operational purposes, while others are charged with the responsibilty of providing (or assisting or ensuring the provision of) such properties as housing, particularly housing for the less well-off members of society. In many countries these powers are referred to as the power to "resume interests in land", and the process of implementing the powers is known as "resumption".

Even when such powers exist, the compulsory acquisition of land for development purposes is not without problems as:

(i)   Land can only be acquired for purposes permitted under the powers granted to the acquiring authority, and these may not include making land available for private development, or they may restrict the type of development for which land can be acquired.

(ii)   Completion of the necessary acquisition procedures may be time-consuming and costly as acquisition orders will normally be subject to specific rules relating to their publicity and confirmation, to possible appeals, and to their eventual implementation if they are confirmed. Additionally, disputes as to compensation may occur, which in themselves may be costly, and which may lead to uncertainty as to the eventual cost of acquisition.

(iii)   Compulsory acquisition may frequently lead to appeal or court proceedings regarding both the actual acquisition and compensation for the interest acquired. These proceedings are likely to cause delay and to further increase the total cost of compulsory acquisition.

(iv)   Where acquisition is based on the assumption that normal market value must be paid for each interest in land acquired, acquisition costs may result for a scheme such that the total compensation paid for all the necessary interests actually exceeds the total development value of the assembled site. It may easily happen that the total value of all the parts actually exceeds the value of the whole, in which case the acquiring authority which has assembled the development site may be in the position of having to sell the assembled site at a loss if private sector redevelopment of the site is to be achieved. If the site is not offered to the private sector or fails to sell when offered, the authority will itself be faced with the task of redeveloping and bearing the loss on site acquisition.

Consequently, even where a public authority has the power to compulsorily acquire land, the process of the assembly of suitable development sites is not necessarily straightforward, neither is it necessarily quick or possible within acceptable cost limits. Costs will normally need to be low enough to ensure that a sufficiently large development profit is likely to exist which will both encourage and justify the redevelopment of the properties.

In the period after the Second World War there was a considerable amount of site assembly of this nature undertaken in the United Kingdom, much of it involving the acquisition of old, sub-standard, intensively developed housing in areas which had often become slums. It is highly likely that if financial analysis of many of those redevelopment schemes was undertaken it would be found that in development appraisal terms a loss was made, but it should not be forgotten that such money losses could in fact be argued as reasonable in view of the social benefits of replacing unfit housing with housing considered appropriate for healthy and safe modern living.

In the 1990s in particular there has been a trend in many countries towards less public sector activity and the encouragement of private sector initiatives in areas which for many years had been considered to be the province of the public sector only. There are a variety of reasons for this trend, including the fact that many believe the public sector to have been inefficient and too costly in many of its past undertakings. There is little doubt that some governments have seen such a trend as part and parcel of the process of reducing taxation, this being attractive to them as it is likely to increase their popularity at the ballot box, the reduction of public sector activity resulting in reduced public sector financial needs.

With this trend there are likely to be fewer instances of public sector site assembly through compulsory purchase, but the problems of site assembly will remain and will be transferred to the private sector with the transfer of work to them. It is likely to be an important part of the "privatization trend" to ensure that private sector organizations which take responsibility for the development of roads, railways, tunnels, airports, housing, and other major works such as the provision of water, gas and electricity, are properly catered for with respect to their site assembly needs if the private sector is to be able to operate efficiently in these areas. If it cannot operate efficiently, then the movement of work to it will not necessarily produce benefits. Balancing the needs of society in general and the private sector operators with those of the individual owners of interests in land will be a delicate process in many respects, and probably more difficult than when much similar legislation was enacted in the last century. It is likely to be more difficult because modern information systems and the generally higher levels of education among the populace of most countries may well result in higher levels of opposition both to the granting of and the exercise of the powers of compulsory acquisition.

Chapter 10

Identifying Demand for Development

There have already been regular references in this book to the need to determine the level of effective demand for an envisaged project before a decision is taken to go ahead, and before the detailed design of a project is determined. The key motive behind most developments being to develop a property which can ultimately be marketed at a profit, in order to maximize that potential profit, a developer will wish to create a property which can be disposed of rapidly, either through rental or sale (or both), at the best possible price.

In the relatively early stages of the development process market research should therefore be undertaken in an attempt to assess the market demand for the particular type of property which it is intended to develop, in the location in which it is proposed to develop and at the time the property is likely to be offered on the market. The latter point is particularly important as there will inevitably be a time lag between forming the development concept and the delivery of the completed development, so the prediction of future demand is more important than the determination of current demand, although the latter is likely to be an important starting point in assessing the future. The research should therefore attempt to identify the key factors which are currently affecting demand, but more importantly those which are likely to influence future demand.

Overall, the developer will wish market research to provide information which helps to ensure that any development undertaken is specifically designed to fit in with demand patterns which are likely to be current at the time of completion. This will reduce the developer's risks and increase the likelihood of successful marketing of the finished product. The possession of good background information should enable a developer to make better informed decisions in order to ensure the likely returns from a scheme more than compensate for the risks entailed, which risks the developer will seek to control and reduce as much as possible.

The views formed as a result of market research can at best only be opinions; they cannot be precise answers, and their accuracy will be affected by the depth of the research and the degree of skill with which it is carried out. They will also be very much affected by how events in the future actually relate to the predictions made at the time the research is undertaken, and it is inevitable that there will be some completely unpredictable developments in most scenarios.

Demand for new properties may result from a large range of factors including population growth; the increased purchasing power of the population; changing age patterns in the population; movements of population; changing consumer preferences; current shortages of supply in the relevant type of property; current inadequacies of quality in the stock of the relevant type of property; changes in technology, in industrial practices, in marketing practices, etc. which render existing properties obsolete; and the need for more appropriately located properties than the current stock provides. In any research exercise determination of the factors which are most influential in creating demand will be very important, as will the identification of any future developments which might alter the impact of those factors and so alter the strength of demand.

The assessment of demand for different types of property is important to private sector developers because their objective is to make profits. It is important to the public sector because of the need to:

(i)   achieve good planning;
(ii)   minimize unneccessary and excessive traffic generation;
(iii)   assist in the supply of an adequate stock and range of all types of property to satisfy the needs of society;
(iv)   minimize urban degradation and in particular inner-city decay;
(v)   supply suitable infrastructures at the appropriate time;
(vi)   take a broad overview about the overall well-being of its planning area;
(vii)   observe all fundamental planning principles; and
(viii) seek to achieve all the objectives of planning and in particular to avoid conflicts between different land uses and different land users.

Whilst the public sector may not be operating in the market-place, it nevertheless serves a society which expresses its demands and preferences through the market, so the public sector can serve society more effectively if it is in possession of good market research with respect to society's property needs.

Additionally, the effect of development proposals on existing infrastructures and on use patterns of the various public utilities and services is important. It is important for the public sector to have dependable background information to enable it to reliably assess development proposals and the assumptions made by developers to support their proposals.

A planning authority's market research will need to take great account of proposals in neighbouring competing centres when assessing individual submissions, for even if individual developers do not adequately consider such matters it is important for the public sector to do so in the interests of trying to achieve the most efficient use of the infrastructure and of services provided by the public sector in general.

In reality, the market research objectives of both developers and local planning authorities should be similar: they should be attempting to make realistic assessments of the patterns of effective demand. Developers want their developments to be successful in order that they will not lose money and that they will in fact make profits, whilst the public sector organizations should not want to have property failures in their areas which are likely to cause problems in the longer term. On the other hand, successful developments will help to increase the overall attractiveness and quality of the area for which they are responsible, and should help to increase returns from local taxes where part of those returns is based on property values, or even if they are based on local income taxes.

Planners and other public sector groups will, however, be interested in the potential movement of property uses from existing properties in a different sense to developers. The latter will be interested in the possibility of existing property users moving to their new development, which would help to make it successful, but they are likely to be indifferent to the effect their development will have on existing properties unless they happen to own those properties. Those in the public sector should, however, consider the likely effect of new development on existing properties very carefully, particularly if the effect is likely to be a movement of users from existing properties leading to the under-use of those properties and the possible deterioration of the areas in which they are located, and also to a changing demand pattern for local services.

A developer wishes market research to indicate the likely level of success of the proposed scheme, but planners will wish market research to indicate the wider implications as well. A developer may propose a new retail development, for example, believing that because of its modernity and convenience there will be sufficient demand both from retailers and shoppers to make it successful. However, overall there may be no existing shortage of retail outlets and the public sector planner should consider the effect such a new development is likely to have on existing retail properties and the area in which they are located.

Market surveys

The starting point should be to *assess the existing situation*, particularly the total amount of space or total number of units which currently exist of the type of accommodation it is proposed to develop. In addition, the amount which is being added to the stock on an annual basis should be calculated, as should the amount which goes out of use each year. The reasons for such space going out of use should also be determined, for this could be a guide as to the type of accommodation which ought to be designed.

The *suitability of the existing stock* should be assessed through a survey of the physical condition of existing property. The economic suitability of the existing stock should also be reviewed; for instance it should be determined whether any of the stock is currently obsolete or obsolescent or is likely to become obsolescent in the relatively near future, and if so the reasons for this occurring should be ascertained.

Wherever possible it is sensible to use *existing sources of information*, but the accuracy and reliability of such information should be checked before undue reliance is placed upon it. The reasons for which it was collected and the basis upon which it was collected should be ascertained, and also the date on which it was collected and the date at which the information applied, for they may be different dates. This may well affect the relevance of any existing information to a new market survey, and dated information should be updated as accurately as possible with note being taken of any subsequent changes in trends or any other changes which might be relevant to the development decision.

If, for example, a retail development is being considered, developments which have been completed since the last figures were compiled should be listed, as should any *developments currently being constructed,* together with those not started but for which planning approval has been given. The expected completion dates of work in progress or yet to be started could also be important information.

Population figures will frequently need updating, but in doing this it will be important to note any changes in the spending power of the population, and also any changes in the ability and the propensity to spend since the last population figures were compiled. It should not automatically be assumed that people will spend in the locality in which they live, and the likely location of their expenditure should also be carefully estimated. The latter may be affected by changes in existing shopping outlets, for instance the refurbishment of and change of retail mix in an existing shopping centre, or by other new developments which are not yet finished. It would be sensible for individual developers to take account of the proposals of other developers otherwise there can too easily be a case in which two or more developers may each be dependent upon the same population group for the success of their new developments, the size of the group in such cases often being insufficient to support all of them.

The objective of market research should not be to justify a development proposal but to assist in the realistic assessment of its likely financial success, or otherwise.

Good market research may well prove to a developer that he or she should not proceed with a development proposal as it is likely to result in financial loss or an inadequate return. Research, which results in such a finding and causes a developer to abandon a scheme for which financial success is doubtful, is research which is well worth doing.

To pursue the example of a possible retail development, a retail *market survey* should be undertaken in the hope of determining many facts including the following:

(i) The geographical limits of a catchment area, which may be determined by such features as breaks caused by hill or mountain ranges, or by the existence of desert or infertile tracts of land.

(ii) The economic catchment area, which may well be smaller than a natural geographic catchment area if there are other competing areas within easy access.

(iii)   The population and trends in population in the defined catchment area.

(iv)   The estimated income levels and the spending power of the catchment area population, together with details of trends in income levels and spending power. The age structure of the population is important, for young people are likely to have different spending patterns to older people, and for similar reasons the marital and family statuses of the population are important. Any trends which indicate changes in these structures are also important, for such changes are likely to result in changes both in the ability to spend and the patterns of expenditure.

(v)   The major sources of employment in the area, together with information regarding unemployment levels and any factors likely to affect the local economy, employment levels, and types of employment. The current state of the local economy and the anticipated future state of that economy are particularly important considerations, as the majority of trade for most centres will be expected to come from the area in which a centre is located. Very careful analysis of all economic indicators should be undertaken.

(vi)   Information as to whether the existing retail outlets adequately cater for the present population of the area, and whether the present situation is likely to continue or to vary in the future.

(vii)   Travel patterns in the area together with information as to the adequacy of local public transport systems and service provision, and information regarding any proposed or likely changes to systems and services. Information regarding the adequacy or otherwise of car parking provision, and information on the current methods of travel and the preferred methods of travel of shoppers should also be sought. Any deficiencies in travel services and facilities should be identified. Likely or possible changes to rail services or to road networks should be carefully researched.

(viii) The demand pattern for retail property as revealed by evidence of past transactions, and the prices paid for retail accommodation both in rental and capital terms. Trends in demand and prices are most important, as also is information on the type of property for which there is still unsatisfied demand. With retail properties the type of retailer most likely to require space is important, as different retail trades have different profit potential, and hence the level of rents which can reasonably be expected to be commanded by a development is likely to vary greatly depending upon which retailers are most likely to take space in the proposed development.

(ix) The current supply of retail space in the catchment area and its quality and suitability in design terms and physical terms.

(x) The degree to which the existing stock is used with details of any level of inadequacy of use or of excess capacity.

(xi) Whether, as indicated by (ix) and (x) above, those owning existing retail areas might find it easy or not to respond to new challenges from new developments, and whether more intensive use can be made of existing facilities.

(xii) Information about why existing shoppers use specific existing retail areas.

(xiii) Information about why infrequent shoppers do not use existing retail areas more frequently.

(xiv) What deficiencies infrequent and existing shoppers identify and what improvements in retail areas they consider to be desirable.

(xv) What deficiencies existing retailers identify in current retail areas and what improvements they would like and would be prepared to pay for.

(xvi) Precise details of proposed new competing developments and other possible new developments, in order to decide whether they will, in overall terms, be beneficial to the proposed development or a threat to it. The research should not be restricted to the immediate catchment area, for developments may be proposed in nearby competing areas which could well draw trade from the catchment area.

(xvii) What type of new development is required if a need is revealed by the survey. In particular information relating to location, the type of development, the quality of construction which would be justified, desirable design features, the type of retailing outlets needed, the size of retail units needed, and the rents likely to be obtained from any new development should be sought.

The above questions and information are particularly relevant to a possible new retail development, and although the questions to be asked for new residential, industrial or commercial development may vary, the overall objectives will be similar. There are no definitive lists of information which should be sought or questions which should be asked for each type of proposed development; the above list is intended to be indicative and to give a general understanding of information which is typically sought through market research. In each specific situation it will be the task of the developer and the researcher to determine exactly what information is likely to be useful to the developer, and to "tailor-make" the research exercise to suit the proposed project.

Sources of information

It was earlier suggested that, whenever possible, existing sources of information should be used, although the reader is reminded that such information has to be utilised with great care. However, when existing information is available, its use is likely to result in savings of both time and cost. Typical sources of useful information are likely to include:

(i)   national census figures, particularly with respect to the population and its make-up;
(ii)   other government statistical records, such as trade figures and details of industrial and business activity;
(iii) locally commissioned surveys, such as those done for local authorities or other local or central government groups, or those commissioned by private sector organizations;
(iv)   statistics collected by taxation or rating authorities;
(v)   reports published by professional firms such as property consultants, accountancy firms, economic consultants, and research groups; and
(vi)   bank and building society reports and surveys.

There will frequently be a considerable amount of information available from such sources and enquiry of local authorities and public libraries will usually reveal its existence. In addition, local estate agents and property consultants will also be a useful source of information particularly with respect to the property market. However, care needs to be taken to ensure that double counting (or worse) does not occur when seeking to estimate market demand by adding together the estimates of a number of different agents. All hearsay evidence of any type should always be treated with extreme caution.

Predicting the future

A major problem with the results of market research is deciding how exactly to use the information. The first task is to try to achieve the greatest possible accuracy in the compilation of survey information, but the interpretation of all information collected is also extremely important. It may be that a decision has to be made to discard some information as unreliable, or to place greater weight on some information than on other information, and the use of careful judgement may well be needed in making such decisions.

Subsequently, predicting the future becomes a part of the exercise as the development process relates to the future rather than to the present, and it is future needs and uses which are of major importance. Current information therefore has to be adjusted to take account of expectations regarding the future. This is an extremely difficult task, events tomorrow being difficult enough to predict accurately let alone having to look even further ahead.

Economic and financial predictions are extremely difficult to make with accuracy, and if one doubts this it is only necessary to consider how frequently governments, with a wide range of experts and other resources at their disposal, get predictions of the future wrong, and how frequently expert commentators who specialize in such matters, differ in their opinions, opinions which are based on similar information and similar circumstances. When making predictions it is probably advisable to make a range of predictions, and to test each prediction by use of sensitivity analysis. The objective will be to determine how vulnerable each projected scenario is to changes in any of the underlying market, economic, financial or other factors. Overall the analyst will seek to determine which projections are the least vulnerable to possible future changes, and which, if any, are the most reliable and most acceptable predictions.

In undertaking these tasks objectivity is very important, but it is probably impossible to rule out some element of subjectivity as this almost inevitably enters into the opinion forming and judgemental processes, and in undertaking a market survey and using the results it is likely to be impossible to avoid these two exercises at some stage.

In any event it is probably impossible to rule subjectivity out altogether, as there is likely to be subjectivity involved in at least some of the survey answers given by those surveyed. Care must therefore be taken to reduce this type of problem by devising question formats which, as much as possible, eliminate subjective answers. The actual compilation of market survey questionnaires is, therefore, a very important if difficult process. It is critical that questionnaires are devised with great care if the results of surveys are to be dependable. It should not be forgotten that the expenditure of very large sums of money and the acceptance of considerable risks may hang on the results of market surveys. The larger the scheme and the greater the amount of finance involved, the more a developer will be advised to spend on thorough market research at an early stage. However, even for small schemes which clearly cannot justify excessive expenditure on such procedures, there is nevertheless no substitute for thorough market research at an early stage if a project is to maximize potential success, or if bad schemes are to be avoided altogether.

Chapter 11

Site Identification and Assessment

The identification of a range of possible sites for a development project and the selection and purchase of the most appropriate site requires thorough investigation and careful consideration, for the selection of a suitable site is crucial if the profit from a project is to be maximized, whilst choice of the wrong site can ruin an otherwise commendable project. Indeed, if a site which fulfils most of the main selection criteria for a particular project cannot be found, it is arguable that a project should be abandoned, particularly if it involves a use which is particularly sensitive to locational considerations and only compromise locations are available. With shops, for example, properties which are similar in physical terms can have very different retailing potential even though their sites may only be about 50 metres apart, while big differences in value can occur between two similar residential properties simply because one property is in a more sought-after residential area.

Where a choice of sites exists, site conditions may have considerable implications for development costs and also for the time required to complete a project. A site which, because of its physical features, is difficult to develop can result in big increases in the cost of construction, additional interest charges on finance because of an extended construction period and, for the same reason, the delayed receipt of returns from the completed project. The cumulative effect of cost increases and the delayed receipt of rent or sale proceeds can be very considerable and, at times when there is a very competitive market situation or a very limited market for a type of property, too much delay may result in a project "missing the market", that is by the time it is ready for use there is no longer any unsatisfied demand at a price which will produce a profit.

The attitude of the local planning authority may be very important in the site selection process. While one site may be as near to ideal as it is possible to get from the developer's viewpoint, it may be that it is not favoured by local planners for the proposed development. As a result planning procedures may become lengthy and possibly unproductive if approval is not ultimately given, or if approval is only given for a project which is not financially attractive. Planning negotiations, the application process, and the appeal process if it is necessary, can be both very time-consuming and costly, whilst if conditions are placed on an approval which are likely to be costly to comply with, or if an approval only allows inadequate development with the result that an acceptable profit is unlikely to be made, what appeared to be the ideal site may in the event prove to be far from ideal. It may be that because of planning considerations a developer will sometimes be well advised to choose a slightly inferior site on which the granting of an acceptable approval is likely to be timely, with the result that there is no costly impediment to the development.

Consequently, the ability to compare the merits and the financial evaluations of a number of possible development sites is a great advantage, as it may well be that the financial appraisals indicate that the best site for a development is one which might not initially have been considered to be the best choice. At the end of the day, the site which is likely to provide the best return on the capital investment will be the best to develop if profit maximization is a development objective. If the profit motive does not exist, (as may be the case with some development done by charities or public authorities), the costs of development and anticipated returns, if any, must still be important considerations in site selection, as must the opportunity cost of capital devoted to a project. (The "opportunity cost" is a sacrifice made as a result of taking one course of action rather than an alternative; in this case quantification of the opportunity cost would be the profit which could be achieved from the "next best" investment opportunity.)

In reality, for a variety of reasons, there will frequently not be a choice of suitable sites, and sometimes there may be no suitable site at all. The most likely reasons for a lack of choice are that the locational requirements of a project are so specific that choice is automatically restricted, or that an area is already so highly developed that vacant land or suitable redevelopment sites are not available. If land assembly is likely to be necessary, the availabilty, or lack of, enough suitable land blocks simultaneously may restrict choice, while as previously implied the planning authority may only be prepared to approve certain types of development in very clearly designated locations. It may also be that the requirements of the financial institution from which funds are being received restrict the choice of sites, in that they may only be prepared to lend for development on sites which they regard as being particularly safe, and their criteria for site assessment may not necessarily be the same as a developer's.

Some of the factors likely to be very important in determining the suitability of a site have already been briefly considered in Chapter 8, and those matters will now be further considered.

Probably the most important consideration in site selection for any form of development is the quality of *location*. Indeed, many property people say that the three most important considerations are "location, location and location", so critical do many consider it to be. There is no doubt that the quality of a site's location is regularly the most important factor in determining the value of a property, although the factors which determine the quality of location may be different depending upon the type of development.

Warehouse and industrial development sites need to be well served by communications networks, especially good quality roads, with rail access being essential for some users, particularly if bulky goods have to be delivered either to or from the site on a regular basis. Some industrial sites need to be located close to adequate supplies of water for use in industrial processes, and there may be other locational criteria, such as the need to be close to airports or docks, depending upon the likely industrial and warehousing activities of occupiers. Accessibility to a pool of suitable workers is also important.

For *retail development* it is important that the location allows easy access for shoppers travelling by a range of forms of transport including car, rail, bus and foot. Easy access for the delivery of goods is also important. Unless a shopping centre development on a "green field site" is involved, it is generally also important for a site to be located in an established shopping area, although there is now a trend for self-contained groups of retail outlets to be developed in out-of-town and edge-of-town locations, often on sites which are being redeveloped. With such developments it is critical that the site is located in an easily accessible location particularly for travel by car, while sites need to be large enough to allow the development of a centre which is big enough to contain a range of outlets sufficiently attractive to draw customers to it. Apart from fully self-contained shopping centres of neighbourhood, subregional or regional size, there are now bulky-goods centres and factory or warehouse shopping complexes being developed in many parts of the world.

The locational requirements for *office developments* again include being easily accessible by road, and if possible by rail also, to allow easy access for office workers and visitors to the offices, and also for those providing back up services to office tenants. It is also a benefit for offices to be located close to retail areas where eating and shopping facilities will be readily available for office workers. In that respect the two types of use are often complementary, and will frequently be located adjacent to each other. Depending on the type of occupier likely to use an office development, it may well be desirable for a site to be located close to complementary office users. This has frequently happened in large cities which may, for example, have banking sectors, insurance sectors, and others including legal sectors. For some office users being located near to other office users may not be critical, but it should not be forgotten that if a user chooses to develop outside an acknowledged commercial centre then the value of the offices may be restricted should the user ever wish to dispose of them.

An important locational requirement for *residential development* will again be for a site to be well served by good roads, whilst it is an additional benefit if there is a convenient rail service. Residential development will also benefit by having easy access to the range of properties which residents are likely to use, such as schools and hospitals, retail areas, and sports, recreational and entertainment areas. Ease of travel to places of work is another desirable feature. Another important locational consideration is that residential development should not be located too close to uses and developments with which it is incompatible, such as heavy industrial uses or intensive agricultural operations which produce objectionable smells.

Communications facilities in the form of road and rail networks have already been referred to quite extensively above and, in assessing the quality of communications serving a particular site, proposed improvements should be considered as well as existing facilities. Planned improvements elsewhere may adversely affect a location and should not be overlooked, as works which will in due course benefit other locations may result in a specific location becoming less attractive to users in relative terms with the passage of time, and accordingly becoming less valuable also.

The importance of the availability of adequate *services* to a development site has already been stressed, and in choosing between a number of possible sites the adequacy of services or otherwise could determine which site is ultimately chosen. In addition to the questions of whether sufficient service provision already exists in the locality in the form of mains water, electric, and sewers in particular, and the cost of connecting these services to the site if they are not already connected, there may be the need to consider whether adequate provision can in fact be made, and how much time will be needed to make the connections when they are required. In considering such matters, it should not be overlooked that building extensions – which would place further demands on services – may be needed at some future date. Neither should the possibility be ignored of more intensive use of services being made in the future simply because of a change of occupier requirements.

The *physical features* of a site will be important in determining whether it is suitable both for a particular type of development and also for a specific development project, and shortcomings may rule out certain types of development completely.

Industrial and warehouse developments nowadays will normally be done in the form of estate developments, and accordingly developers will generally require large areas of land which are either level or which have only moderate changes in levels throughout the site. It will be preferable for each separate industrial or warehouse unit to be on a level site, for changes in levels can lead to inefficient production flows which in turn increase production costs. Development on sharply undulating or steeply sloping sites is likely to be avoided unless no better sites are available or unless units are developed for the users of small industrial units. It will also be desirable with this type of development for sites to have good drainage characteristics otherwise expensive site drainage works may be required, while for heavy industries and large buildings, sites with good sub-soil conditions and good load-bearing qualities will be necessary if very expensive foundation work is to be avoided.

If *large retail centres*, such as shopping centres or factory style outlets are to be developed, the same type of site features will be required as those outlined for industrial and warehouse developments.

Retail development in existing retail areas is likely to leave a developer with little choice other than to accept whatever sites may be available, and if the location is acceptable in economic terms a developer may be persuaded to develop or redevelop irrespective of whether the physical features of the site are ideal or less than ideal. In such circumstances the financial appraisal of the scheme will take into account the fact that expensive foundations may be needed or that the site is less than ideal in terms of shape, size, or physical features, and will thereby indicate whether development on the site is likely to prove profitable.

Office developments vary enormously in size and accordingly the importance of physical site features is likely to vary depending on the specific scheme. For large, multi-storey buildings, good load-bearing characteristics and good site drainage will be advantageous, and both are likely to reduce construction costs considerably in contrast to sites which do not have those good features. However, as with development in existing retail centres, it will often be the case that office development occurs in areas which are already established as office precincts. In this case there may be little or no ability to choose between different sites and a developer is likely to have to select a site "for better or worse", accepting any deficiencies it may have in physical terms and developing if the anticipated value of the completed development is more than adequate to cover any costs which result from site deficiencies.

In the case of the development of *office parks* in suburban or edge-of-town locations, the ideal site characteristics will be similar to those for warehouse and industrial estates. However, the fact that office users will frequently require smaller units of occupation than industrial or warehouse users, and that large delivery vehicles will not use sites to the same degree, does mean that less level and more undulating sites can probably be developed on an economic basis. The changes in level may even be used to provide a more interesting estate environment, particularly if good landscaping is done, which may result in enhanced market values.

Large scale residential development in the form of detached, semi-detached or terraced housing is likely to be easier on reasonably level land, although undulating or sloping land may lend itself to a more interesting overall environment, whilst good drainage characteristics will again be desirable. However, even when such site features are not available, the small scale of most units of accommodation permits development even on steeply sloping sites, while land drainage systems can be reasonably easily installed and different types of foundation can be used to suit varying load-bearing qualities. However, when difficult site conditions exist large-scale residential development is only likely to occur if relatively high market prices are commanded by the finished product, sufficiently high to justify the inevitable higher costs of development.

The site requirements for *high-rise residential development* will be similar to those for multi-storey office developments.

The actual decision as to whether a specific site is thought to be suitable for a particular development project will frequently be based on a number of considerations and will involve the balancing of advantages and disadvantages. For large schemes the decision is likely to depend upon the collective judgements of several members of the development team, such as the valuer, site engineer, architect, and quantity surveyor, although ultimately the decision will have to be made by the person carrying the final responsibility for the project, that is the developer. When there is no choice because only one site is available, it will still be necessary for the individual judgements to be considered in making a decision whether to develop.

Whether it is thought appropriate to proceed with a project will ultimately be determined by the financial appraisal of the project based on the site under consideration. The judgements of the various specialists will be reflected in financial terms through the values assessed for the completed development and in the figures predicted for construction costs and other essential costs. Above all, with most developments, in economic terms the location is likely to be the most important determinant of the viability of a project, as the quality of location will determine exactly what levels of value are likely to be commanded by the completed development in the market place, in terms of both rents and capital value. The financial assessment of sites will be considered in detail in Chapters 13 and 14.

Chapter 12

Finance for Development

Arguably the most critical part of any development is arranging the right "financial package", as projects which are in all other respects good, can go wrong because of inappropriate financial arrangements, whilst what appear to be poorer projects can be successful as a result of appropriate financial arrangements being made at the outset. Most modern development projects require substantial sums of money for their implementation, and this invariably creates the need for borrowed money if a project is to proceed.

Equity and debt funding

Projects can be funded through the use of *"equity"*, that is a developer's own money, or through *"debt" funding*, that is through money borrowed from a third party such as a bank or other financial institution. The most common arrangement is probably for a combination of "equity" and "debt" funding to be used.

If a developer has sufficient cash available there are some advantages in using equity to fund a project, including:

(i)   as long as the developer remains solvent, there will be certainty of funding and no dependence upon another organization for finance;

(ii)   the cost of finance for the development will be known throughout the scheme, being the organization's own required rate of return on funds, or the opportunity cost of those funds;

(iii)   there will be no onerous conditions or terms imposed by a third party and no payment of fees to another organization; and

(iv)   overall financial control of the project remains with the developer.

On the other hand, there are disadvantages in using equity, which include:

(a)   the organization's own money is put at risk when undertaking the project;

(b)   the benefits of favourable gearing (the relationship of equity to debt in the overall financial package, which will be covered later in the chapter) will not be available;

(c)   if too much equity is used in one project, the number of projects which can be undertaken simultaneously is reduced;

(d)   depending upon the tax regulations in the country in which the development occurs, the ability to claim tax relief may be lost, although this will not necessarily be the case in many countries.

The use of debt funding also has advantages, which include:

(1)   depending upon the terms of the loan, the risks inherent in the project may be transferred to another organization's money, although this will also depend upon the debt to equity ratio of the overall financial package for the project;

(2)   all costs of borrowing can usually be set against tax, so reducing the burden of those costs;

(3)   by the use of appropriate gearing, profits can be made using someone else's money;

(4)   using debt finance to fund a project releases equity for use in other projects, and if suitable gearing is arranged for each project and they are all sound projects, multiple profits can be made on the equity element, those profits being enhanced by additional profits made on someone else's money, that is the debt funding. Whilst there may be risks in such an arrangement, it may also assist risk reduction in that carrying out a number of projects enables risks to be spread between those projects, rather than all the risks being borne by one project. It may also assist in reducing risk by allowing diversification into the development of different types of property, rather than all of the risk being borne by one property type.

There are, however, disadvantages in using debt funding, which include:

(i)   the borrower of funds will almost inevitably lose some control over the project to the lender of the funds;

(ii)   there may not be such certainty of funding in terms of the amount available, the period for which it is available, and the cost of those funds as there would be with equity funding. In addition, if loan terms are agreed which provide certainty in those areas there will almost inevitably be a higher charge made for the funds, and the borrower is again likely to lose some control of the project to the lender;

(iii) it may be a fundamental condition of some loans that the lender is allowed equity participation, that is the lender has the right to contribute some equity to the project and to have a share of the profits produced by the development;

(iv) having revealed details of the project to a possible lender, the developer may find that the lender proceeds to undertake the project itself, ignoring the developer completely.

The decision whether to borrow funds or not is therefore very important, but in reality it is a decision over which, in many instances, the developer is unlikely to have any choice, borrowed funds regularly being essential if a project is to be undertaken. That being the case, it becomes important to decide how much should be borrowed and how much equity funding should be relied upon. This may in turn determine whether a project proceeds or is abandoned, the accountability aspects of the project (that is the relative responsibilities for the risks inherent in the project), the size of the project, and the distribution of any profits which may in due course be made.

The developer's requirements

From the outset of a project a developer needs to be certain of having a sufficiently large loan for a guaranteed time period which will allow sufficient time for the design, construction, leasing and marketing of the project, and which also has a contingency period built in to cater for any delays which may occur. A developer will also wish to have a definite rate of interest fixed for the period of the loan, and will not wish to have a loan with an interest rate which may vary, unless it is in a downward direction only (such an arrangement being extremely unlikely to be available in practice), because if interest rates should rise the cost of a project will increase. In many instances it will be preferable to a developer to pay a higher rate of interest at a fixed rate than to take a loan with a lower rate of interest knowing that it may well increase during the loan period. The developer will also wish to be certain of having funds available for the development period, and will resist getting into a situation in which loan funds might be recalled by the lender at an inconvenient time. In addition, a developer will seek to have as much flexibility as possible in his or her favour in other loan terms, such as break clauses, repayment clauses, and provisions for refinancing.

At an early stage a developer will need to decide whether only *"short-term finance"* is required or whether *"long-term finance"* is also needed. Short-term development funding is made available for the development period only, having to be repaid following the completion or sale of the project, and it is therefore only secured upon the value of the development site, the development concept, and the construction work as it proceeds. Long-term funding will be required if a development is to be retained following completion of the project, and will be necessary if the developer wishes to retain the project as an investment. On the assumption that the development when finished is successfully let to tenants, long-term funding will be secured upon a completed and income-producing property, and this should present less risk to a funding institution than lending on a development site and incomplete project for which the ultimate success is not certain. For this reason short-term funding will almost certainly entail a higher interest rate than long-term funding, and there is likely to be a 2% to 3% gap between the cost of the two depending upon many variables such as the risk inherent in the project and the reputation of the borrower.

As short-term funding is relatively expensive, and as a developer will in any event wish to contain costs as much as possible, he or she will wish to have access to it on a "draw-down" basis, that is to have an arrangement that the total sum required will be made available but will only be used as and when needed. The developer is therefore likely initially to use money for the purchase of the site and associated costs, and then at a later date to use funds for design costs, and for other payments only when required, so minimizing the amount borrowed at any time and thereby loan interest charges as well.

If the developer also makes arrangements for long-term funding, he or she will wish to have an adequately long loan period agreed to enable the capital sum borrowed to be paid back in sufficiently small annual payments to ensure that repayment of the loan will be relatively painless. It is desirable that the income produced by the completed development should be large enough to cover loan interest charges and the capital repayment, while at the same time leaving a further sum available for the developer to provide an adequate return to equity plus a profit. The loan period should therefore be negotiated with those requirements in mind.

The lender's requirements

The lender will in fact be lending money to someone or some organization which hopes to make a profit from the borrowed funds, and the interest charged by the lender will represent payment for the use of those funds. The lending organization is in the position that, unless it becomes an equity partner in the scheme with the right to share any profits, the only returns which it will receive will be the interest paid by the lender together with any fees associated with the loan, such as arrangement fees and loan management charges.

The lender therefore is becoming involved with a risky undertaking in the form of a development project, and will wish to ensure an adequate return on the funds lent for the duration of the loan, plus the return of the sum lent at the end of the loan period. As indicated earlier, during the development period the loan funds will only be secured on the development concept, the site value, and the amount of construction work completed at any one time. These may be insufficient security to cover the loan funds outstanding at any point in time, and in fact should the scheme fail leaving construction work partially completed, the total value might in some circumstances be lower than if the site was clear and ready for development, the partially completed works sometimes being a liability rather than an asset. *"Security"* or *"collateral"*, in addition to the security offered by the project, is therefore likely to be sought for the duration of the loan period. A lender is likely to require security over the scheme itself, plus, if possible, security over other properties owned by the borrower. If the borrower cannot provide other security the lender is likely to require guarantees from other parties, in which circumstances the reputation and financial standing of those other parties will be important.

It is suggested that, above all other considerations, the competence, reputation, integrity and financial standing of the developer are probably the most important considerations from the lender's viewpoint, as even a good project may be a very big risk if the borrower is either incompetent or lacking in integrity or honesty, while marginal schemes may be made successful if implemented by a highly skilled developer. Additionally, even when schemes are not as successful as originally hoped for, if the borrower, be they an individual or an organization, has high integrity the loan funds are nevertheless likely to be more secure than if the borrower has little or no integrity.

As already observed, the lender is entering a high-risk area when lending on development projects, and should therefore take every opportunity to minimize the inevitable risks. It should not be forgotten that the lender runs the risk of losing all the funds lent if a project should go wrong and adequate and dependable security has not been provided by the borrower. The lender also runs the risk of becoming the owner of a partially completed development if things go wrong and foreclosure of the loan results in the ownership of the land passing to the lender. This is a situation which most lenders will wish to avoid as they will normally be financial institutions which are not set up to own and manage property. Even if property ownership is part of their normal range of activities, they will usually wish to purchase property investments to fit the policy requirements of deliberate business plans, rather than to acquire properties haphazardly as a result of failed loans.

The lender should initally seek to reduce risk by carefully assessing the would-be borrower, checking on the developer's previous "track record" in terms of development experience and especially experience in the type of project currently proposed . The skill in managing projects on the part of the developer and the building contractor, when appointed, should also be checked, as the ability to work to projected deadlines is important in controlling project costs. The general reputation of the developer should also be carefully checked, together with a full review of his or her financial standing and financial record to date.

The lender should also carefully assess the development scheme itself, very carefully checking the developer's feasibility study and economic and market assessments. Should the lender doubt the reliability of the developer's studies, independent studies should be taken which the lender will almost inevitably require the developer to pay for. Feasibility studies are wide ranging studies of all aspects of a proposed development; the objective of a study is to determine whether a proposal is practical in physical terms and in particular whether it is likely to be financially successful. Chapter 18 deals in depth with this topic.

In considering the overall feasibility of a scheme, a lender should carefully review the quality of location of the project, the financial appraisal and the underlying assumptions either explicit or implicit in that appraisal, and the design aspects of the project. A well-designed scheme is more likely to be financially successful than one which has design shortcomings, and the proposed quality of construction work is important for similar reasons. In considering the construction work, the lender should carefully review the contract arrangements and the proposed relative accountability under the contract of the developer, the various consultants, the builder and the banker.

It would be sensible for a lender to work on the basis that, with the passage of time it could become the property owner if things go wrong, and it would therefore also be sensible for a lender to require contract terms and a quality of design and construction which it would consider appropriate were it undertaking the development project itself.

The lender's funds will be at risk during the planning stage if the required use is not permitted, if floor/space ratios are inadequate (meaning that the required amount of space cannot be developed), if onerous conditions relating to design, construction or the use of the developed property are placed on a planning approval, if there are problems in complying with building code requirements, or if no approval at all is forthcoming either with respect to planning or construction. The realism of the scheme in planning control terms should therefore be carefully considered, while it would be wise for a lender to provide only limited funds, or not to lend any funds at all, until the required approvals have been obtained, so leaving only the developer's funds at risk during the planning stage.

During the construction stage the lender's funds may be placed at risk if construction costs increase because of such things as increases in the cost of materials, inclement weather conditions which delay progress, adverse labour market conditions such as a shortage of skilled tradesmen, industrial disputes, or wage rate increases, and unexpected construction problems such as the unanticipated discovery of difficult sub-soil conditions. The lender should therefore carefully consider the builder's record and skill at dealing with such problems, whilst the overall quality of the work previously done by the builder should be an important consideration also. The realism of the figures actually used in the financial appraisal for these various items should be reviewed, as should the allowance of any contingency sums to cover unexpected costs should they arise.

The lender's funds may also be put at risk if the development is not properly marketed with the result that rents are not produced or a sale is not achieved by the projected dates, or that the rents and sale figure produced are lower than expected. The lender should therefore investigate whether any of the property is subject to "pre-letting" and, if this is the case, the quality of the lessees involved should be investigated, as should the terms of the pre-let arrangements. Similarly, if a sale of the property has been agreed before construction starts, the quality of the purchaser should be assessed as should the terms of the agreement with the developer, for if either intended lessees or an intended purchaser can easily withdraw from their arrangements, the overall security of the project is not increased to a significant extent over the situation which would exist without pre-lets or a pre-sale.

The realism of projected rents and capital values should be carefully considered by a lender, who would be well-advised to ensure that loan funds advanced are likely to be covered even if a shortfall should occur in either.

Even if a lender is only providing short-term funds, it is suggested that it would be wise to check on the long-term funding arrangements and terms. If they are in any way deficient, or not completely assured, and problems caused long-term funding not to eventuate, the short-term funder could again be left owning a property for which it only ever intended to be a lender rather than an owner.

With respect to the lender of long-term funds, it is suggested that all the considerations relevant to short-term lenders remain important to the long-term lender because they all relate to the likely success or failure of a project, while it is absolutely essential that the long-term lender assesses a project from the viewpoint of a long-term investor. A long-term loan is in fact a long-term investment in a project without, however, the possibility of making a long-term capital gain unless an equity participation in the project is arranged. The long-term lender is therefore in a position of not being able to benefit from gains in the capital value of a property, except that loan risk may be reduced as a result of capital appreciation, while, on the other hand, they may well suffer considerably if there are losses in capital value.

It is suggested that it is therefore extremely important for the long-term lender to pay particular regard to the quality of the tenants to whom the developer agrees to let accommodation, and that the leasing terms should be carefully inspected. The long-term lender would be justified in wishing to have an input to the design process for the development, and a say in such things as the quality of materials to be used, for such considerations are likely to affect the long-term value of the property and the long-term running costs. It may be especially important for the long-term lender to consider such matters, as a developer may be tempted to improve the potential short-term profit by using cheaper materials and cheaper forms of construction, which may not be the best long-term policy and which might therefore increase risk to the long-term loan.

Ideally, all lenders of funds for property development should seek only to lend to prime-quality borrowers undertaking developments in prime-quality locations, working to a prime-quality design and construction quality, and using prime-quality consultants and a prime-quality construction firm. Such a combination of circumstances will not often occur and if money was only lent on such a basis many development schemes might not "get off the ground". However, it should be the objective of a lender to identify at the outset ways in which schemes and arrangements depart from the ideal and to make sure that everything possible is done to mitigate the resultant increased risk, and that the loan charges and other loan terms also take account of the specific features of the relevant project.

Many schemes were undertaken in the past on the basis of loan funding of as much as 90% of the total projected development costs, and there is little doubt that, as the result of the acceptance by some lending institutions of extremely optimistic development appraisals by some developers, some schemes actually proceeded on the basis of loan funds which were in excess of the likely completed development value of the project. With some of these schemes the developers themselves were able to proceed by putting very little equity, if any, into the schemes; the result of this being that whilst they stood to make profits if the schemes were successful, they had little to lose if the schemes were unsuccessful as the funds primarily at risk were loan funds. In such circumstances developers may be tempted to take risks which they would not undertake if substantial equity funding were involved, in the latter case the first losses should a scheme go wrong being borne by the developer's equity. It is therefore wise for any lending institution to insist upon a developer having a substantial equity stake in any development project, sufficiently large a stake to ensure that the developer will not take unreasonable risks knowing that the first to suffer from any foolishness will be the developer himself or herself. Where the developer is an organization rather than an individual such a course of action on the part of the lender may not have quite the same effect, as those who will suffer from developments which go wrong will not necessarily be those directly responsible for them, but the shareholders of the organizations who will almost certainly have no direct control over a development. Nevertheless, it is always wise on the part of a lender to insist upon there being a substantial developer's equity in the scheme not only to try to ensure a reasonable approach from the developer, but also to absorb the first element of any losses (and hopefully all losses) which may be incurred.

Because development organizations vary enormously, because each development scheme is unique, and because underlying economic conditions vary over time, it is likely that there will be considerable variations in the details of loan arrangements from scheme to scheme and with the passage of time. Above all the lender will seek to ensure an adequate return is received for the loan of funds, that all interest payments are made by the borrower as and when due, and that the repayment of the capital sum is ensured at the end of the loan period.

"Offshore" borrowing

As interest rates regularly vary from country to country, it may often be the case that loan funds can be obtained at lower rates of interest from overseas sources than from local sources. There is nothing necessarily wrong in borrowing overseas for development purposes, but if this is done the borrower should be aware that, in addition to the normal development risks, the project may also be subjected to the risk of exchange-rate variations.

It is easy to illustrate the type of problem which can result by reference to exchange rates between the Australian dollar and the UK pound, it being quite normal for the rate to vary within a six-month period between a situation in which $1 buys 50 pence and one in which it buys only 40 pence.

With such variations, it could occur that an Australia developer might borrow $15 million from the UK because of lower interest charges, the loan in fact representing £7.5 million at an exchange rate of $1 = 50 pence. The lending institution will require the repayment of £7.5 million (over and above regular interest payments) and, if at the date of repayment $1 only buys 40 pence, the cost of repaying the loan will in fact be $18,750,000. In such a situation the developer will have received loan funds worth $15 million, which will have cost normal interest payments (albeit apparently at an attractive rate), plus a capital loss of $3,750,000 resulting from adverse movements in exchange rates. It should not be overlooked that the adverse movement in exchange rates may in reality have resulted in the rate of interest charged on the loan also becoming higher in real terms than originally anticipated.

If overseas funding is used a developer should therefore seek to minimize the risk of exchange-rate variations either by borrowing, if it is possible, at both a fixed rate of interest and a fixed-exchange rate, so ensuring that the burden of interest payments and capital repayments cannot increase above the figures originally anticipated. Alternatively, it may be possible to take out insurance cover against such risks. If the first arrangement is possible, the cost of borrowing is likely to be increased to cover the fact that risk has been transferred to the overseas lender, while if insurance cover is taken out that also is likely to be costly.

This in no way means that overseas funding should not be used, but it does make it essential that the additional risks are recognised by the developer, that those risks are minimized as much as possible, and that the additional costs which are likely to result are incorporated in all financial assessments of a project.

Gearing (or Leverage)

Reference was made earlier to the "gearing" (sometimes referred to as "leverage") of a project, which was said to indicate the relationship of equity to debt. When a developer borrows most of the required funds and only puts in a small amount of equity a development is said to be "highly geared", whereas if only a small proportion of the total funds is borrowed, a development is described as "lowly geared". A project which is 60% geared uses 60% borrowed money and 40% equity, and so on. The gearing of a project and the rate of interest charged on loan funds can together make a very big difference to the profitability of a venture to a developer, and they can also have a big effect on the allocation of risk between the developer and the lending institution unless the latter takes other steps to ensure that its risk is limited.

The effect of different gearing arrangements can best be illustrated by use of simple examples. For those who live in countries other than the United Kingdom the principles behind the examples remain the same and, if wished, readers can substitute their own currency figures in the examples as appropriate.

Example 1

Suppose a property is to be developed at a total cost of £3,000,000 and the developer expects it to produce a net income of £450,000 per annum when completed. The yield of the completed development in such a case will be 15%, and if the developer completes the development using equity funds, the yield on equity will be 15%.

Example 2

However, if the developer is able to borrow funds at 10% and decides to borrow £1,500,000 the situation will change as below:

Net income from completed development	£450,000
Less	
Loan interest @ 10% pa on £1,500,000	£150,000
Return to developer's equity	£300,000

If the developer borrowed £1,500,000 he or she must have contributed £1,500,000 of equity to the total costs of £3,000,000, and a return of £300,000 represents a yield of 20% from that equity. By borrowing half of the required funds the developer has increased the yield on equity from 15% to 20%.

It should be noted that for the purpose of illustrating the effect of gearing, other considerations such as the effect of tax, the repayment of capital, and the payment of service or management fees have been ignored in these examples.

Example 3

If the developer decided to borrow £2,000,000 of the required funds the situation would be as below:

Net income from completed development	£450,000
Less	
Loan interest @ 10% pa on £2,000,000	£200,000
Return to developer's equity	£250,000

In this example the developer receives an annual return of £250,000 from an equity investment of £1,000,000, this representing a 25% return to equity.

The general situation illustrated by these examples is that *if the rate of return from an investment exceeds the rate of interest charged on loan funds, a developer or investor can increase the yield on equity by increasing the gearing of the project or investment.* Should rents in such a situation turn out to be higher than anticipated the comparisons are even more interesting, and the above simple examples can be revised to take account of the fact that before the development is completed, market rents increase to be 25% higher than the developer originally expected them to be.

Example 4

Had the developer completed the entire scheme with equity funds the cost would have been £3,000,000 and the net annual income of £562,500 would provide a yield of 18.75% on equity.

Example 5

With borrowed funds of £1,500,000 the situation would be:

Net income from completed development	£562,500
Less	
Loan interest @ 10% pa on £1,500,000	£150,000
Return to developer's equity	£412,500

The return provides a yield of 27.5% on the equity of £1,500,000

Example 6

With loan funding of £2,000,000 the situation would be:

Net income from completed development	£562,500
Less	
Loan interest @ 10% pa on £2,000,000	£200,000
Return to developer's equity	£362,500

With equity of £1,000,000 this represents a yield of 36.25% on the developer's equity investment, this being from a development which was originally expected to yield 15% on costs and which in the event yields 18.75% on total costs.

Example 7

If the income produced from the project proved to be even higher the variations in the yield on the developer's equity would be even more pronounced, and if for example income actually proved to be double that originally anticipated, the return on total costs would be £900,000 on £3,000,000 or 30%, but if £2,000,000 of those costs were borrowed at 10% per annum the yield to the developer's equity of £1,000,000 would be 70%.

Situations such as this have occurred in markets in the past with the result that developers have found their equity investments have produced yields far higher than originally anticipated, sometimes producing great wealth in short periods of time. As indicated above, this type of situation may occur when the yield on a development is higher than the cost of borrowing funds, and the higher the gearing in such cases the higher will be the yield on equity.

Example 8

However, there have been situations in which circumstances have been less favourable to developers with interest rates rising quite substantially in situations in which developers did not have loans at fixed rates of interest. In circumstances in which loan rates increased to 20% per annum, the development yield would in the first example have remained at 15%, the net income of £450,000 being the return to the developer's equity of £3,000,000 which the project cost, no loan funds being involved.

Example 9

With borrowed funds of £1,500,000 the situation would be:

Net income from completed development	£450,000
Less	
Loan interest @ 20% pa on £1,500,000	£300,000
Return to developer's equity	£150,000

With equity of £1,500,000 this represents a yield of 10% per annum, or 5% lower than the development yield.

Example 10

With borrowed funds of £2,000,000 the situation would be:

Net income from completed development	£450,000
Less	
Loan interest @ 20% pa on £2,000,000	£400,000
Return to developer's equity	£50,000

With equity invested of £1,000,000 this represents a yield of 5%, which is very much below the yield of 15% on total costs, and which is a dangerously low yield on equity in terms of the risk which would be involved in most development projects. In most circumstances there would be few schemes undertaken for a yield on equity of only 5%.

It can therefore be seen that when the rate of interest charged on borrowed funds exceeds the overall yield on costs there will be a lower yield on equity the higher the level of gearing. It follows that the higher the gearing in such circumstances the more vulnerable will a project be to further increases in the cost of borrowing, while if returns prove to be lower than originally hoped for the situation will be even worse.

Example 11

Suppose that in the event the originally predicted net income of £450,000 was not achieved, net rents being only £380,000, while the rate of interest on loan funds of £2,000,000 turns out to be 20% per annum because of adverse economic conditions:

Net income from completed development	£380,000
Less	
Loan interest @ 20% pa on £2,000,000	£400,000
Loss per annum	£20,000

This type of situation was common in many countries in the major property slumps of the early 1970s and the late 1980s and early 1990s. Developers who were highly geared were faced with increasing interest rates and lower rents than they had hoped for, and many were in the type of situation indicated in Example 11, but with far bigger sums of money involved. Such situations can only be survived for limited periods or if there are substantial other assets and income flows to underpin the losses on new developments, and many developers were not in that fortunate position and consequently "went to the wall".

The general rules about gearing are therefore that if yields from developments exceed the rate of interest on borrowed funds, the rate of return on equity can be increased by gearing and the higher the gearing the higher will be the rate of return on equity funds. On the other hand, if the rate of interest on borrowed funds exceeds the yield from a development, gearing will reduce the yield on equity and the higher the rate of gearing the lower will be the yield on equity.

Developers will, as already indicated, generally have to borrow money to undertake developments, and it is therefore important that they should seek to borrow funds at fixed rates of interest particularly if there is the possibility of increases in interest rates. They should also seek to ensure that they are not so highly geared that they are particularly vulnerable to possible shortfalls in income. If they can only borrow at variable rates of interest they should be particularly cautious with gearing ratios, for it is frequently the case that increases in interest rates will be accompanied by falls in rental levels as both tend to result from adverse economic conditions.

The other side of the coin is that should the developer not be able to borrow at a fixed rate of interest, if the economy improves and interest rates fall and increases in rent occur, then the return to equity will be very much improved and high gearing may result in incredible rates of return to equity. However, any developer contemplating a project with high gearing should be extremely careful to test the sensitivity of the project to increases in interest rates and to shortfalls in income, and one would in any event expect lending institutions to apply the same tests to such projects.

Overall, both a developer and a lending institution should be seeking to ensure that the level of gearing on any development is such that it will enable the development and the developer to weather any forseeable storms, particularly when economic trends are likely to be adverse.

Joint development schemes

The high level of risk inherent in property development, the large sums of money required for many schemes, the increasing size of many modern schemes, and the need to control risk by mitigation or, if possible, the elimination of some risks, have regularly been referred to in this book. Many of these issues can be addressed, in part at least, by joint development schemes in which there is more than one organization acting as developer, such schemes often involving the developer and a funding organization working together.

The trend towards such schemes developed in the 1960s and the major reason for the trend at that time was probably not the desire to share risk, but the desire on the part of funding institutions to participate in the large profits which they saw many developers making through the use of funds they had lent to those developers, often at relatively low rates of interest. Realizing that they held the key to many development projects which could not have proceeded had loan funds not been available, lending institutions were able to insist upon equity sharing arrangements under which, in return for providing some of the funds required to enable the development to proceed, the funding institution received a share of the development profits.

Such arrangements would often also result in a sharing of the development risks as well as any profits. In current conditions, some development projects are so enormous that they would never be undertaken unless the combined expertise of a number of groups was used, but, more particularly, unless a number of groups were prepared to share the risks of such schemes between them, the risks often being too great for one organization alone to bear.

The sheer size both in engineering and financial terms of the Channel Tunnel project to link France and England under the English Channel necessitated the pooling of resources of a number of financial and engineering organizations from both England and France. The financial requirements for such a project were extremely large as also were the inherent risks. Consequently, the pooling of financial resources was needed to raise the necessary capital while such large risks could only be accepted by a consortium in which the risks were shared by members. The "blow-out" in costs which actually occurred in completing the scheme proved the need for a risk-sharing set up.

A similar situation existed with arrangements for the construction of the main Olympic Stadium in Sydney for the year 2000 Olympic Games, and with total estimated costs approaching Australian $400,000,000 and returns which must be extremely difficult to predict, a number of syndicates were set up to submit tenders for the scheme. Such an arrangement means that each party involved does not have to over-commit their resources to the one scheme nor do they have to accept the entire risk of that one scheme. Each syndicate member can continue to use their resources in a range of other schemes without becoming over-exposed to the one venture, and should that high-risk venture for any reason not work out as well as predicted, any problems will be shared between members of the syndicate, each of which will hopefully be prospering on other schemes.

Chapter 13

Development Appraisal

Two methods of valuation are commonly used by valuers for the appraisal of development sites, namely the Method of Comparison, and the Residual Method (which is also known as the Hypothetical Development Valuation Method). For those not familiar with the concepts underlying these methods and the basic principles relating to their application, reference can be made to *An Introduction to Property Valuation* (The Estates Gazette Limited, London, 1994).

The objective of valuation for development purposes is to determine whether development value exists in a site or a property, or alternatively whether development or redevelopment would be unprofitable in financial terms.

In assessing whether development value exists, a valuation should always be done to determine the value of the property concerned for its existing use as development (or redevelopment) is only justified in financial terms if the value of the completed development exceeds the value of the property for its existing use plus the total costs of creating the new development. In carrying out a development there are two main cost items: the current use value of the property which is sacrificed for development purposes, plus all the costs incurred in creating the new development. If the value of the newly developed property is less than the total of those items, the developer will be worse off financially after the development has been undertaken, so only if the estimated value of a new development is at least equal to those two costs should a development project be contemplated.

The objective of the exercise is therefore to reveal whether there is *latent value* in a site or property which can be released by development or redevelopment which will generally entail the expenditure of money on construction work. Sometimes, however, it may merely entail a change in the use of a property for which there may only be limited expenditure required. This would include the cost of obtaining any necessary planning approvals.

The rationale of the Residual Valuation is very simple, namely:

The value of the completed development
Less   The total cost of development *plus* the development profit
Equals   The value of the existing property for development
             purposes

A developer's profit should be included otherwise there will be an implicit assumption that a developer will undertake a development for no financial benefit. In the vast majority of cases there will be no incentive at all to undertake development unless a profit can be made, especially when the large amount of work and the high level of expertise required for most successful developments is taken into account and also when the great risks entailed in such entrepreneurial activities are considered. In a limited number of circumstances developments may be undertaken without the expectation of a profit on the development activity, such situations possibly including development for charitable purposes, development undertaken in pursuit of the corporate activities and objectives of an organization, or because anticipated trading profits or economies resulting from the subsequent use of the developed property may justify doing a development without making a development profit.

The above formula can be varied in format for different purposes as below:

The value of the completed development
Less   The cost of the site
Equals   The sum available to cover all development costs
             plus
             The development profit

Or:

The cost of the site
Plus   The other costs of development
Plus   The required development profit
Equals   The minimum required market value of the completed
             development property

The logic of the basic development valuation is very simple and it forms the basis from which all development decisions should flow.

When used in practice the Residual Method entails making a large number of predictions and estimates as will be seen in examples which follow. Because of the large number of variables which are generally used, the method comes in for a lot of criticism from many quarters, and in particular it is often criticised when it is used in court in evidence to try to establish the value of a property with development potential.

The estimation of the value of the completed development property is itself a difficult task as it involves estimating the value of a building or buildings which do not as yet exist, and which may not in fact be completed and ready for use for many years to come. So large are many modern developments that it is quite common for properties to take five years to complete from the initial conception of a scheme, and predicting rental values, market yields and capital values five years in the future is a highly speculative task. The level of such estimates will depend very much upon the opinions of the valuer involved and in such circumstances it is not uncommon for the opinions of different valuers to vary considerably.

In estimating the total costs of a development project there is a large number of separate items to be costed and a typical project might involve the estimation of the following costs itemised under different cost categories.

Site purchase costs
(i)   Land cost – a market asking price may be quoted for the relevant land or this may in fact be the "residual" figure found after all other costs and values have been estimated.
(ii)   Legal fees and costs related to the land purchase.
(iii) Valuation fees and costs related to the land purchase.
(iv)   The interest charged on the money borrowed for the three items above from the date of expenditure for the entire development period, as no return to this expenditure will be received until the completed development is either let or sold. If equity funding is used for these costs, then the cost to the developer could be included as the opportunity cost of the equity funds. This same approach could be adopted with respect to equity funds used for any other cost items.

Building costs
(v)   Site works, being the costs incurred in such work as demolition of existing buildings, site levelling or filling, subsoil tests, shoring and fencing etc.
(vi)   Costs of road works and the installation of services and drainage works.

(vii)   Cost of constructing buildings etc.

(viii) Cost of landscaping work.

(ix)   Interest charges on all sums of money devoted to all items of building work from the date they will be incurred to the anticipated date when proceeds from the completed development will be received, either as rents or through a sale.

Professional fees and expenses

(x)   The cost of hiring necessary consultants to design and complete the development, which might include architects, quantity surveyors, site engineers, a range of other specialists including planners, electrical engineers, communications specialists, heating and ventilation engineers, interior designers, lift engineers, landscaping consultants, etc. With any other than very small projects a project manager is also likely to be needed.

(xi)   Interest charges on all funds used to pay consultants, such charges being calculated from the date the costs are likely to be incurred to the anticipated date when proceeds from the development are first likely to be received.

Costs of disposal

(xii)   These may include the fees of marketing consultants, solicitors, real estate agents and advertising and promotional costs. These fees relate both to the letting of properties developed for investment purposes and the sale of properties which are not to be retained by a developer.

(xiii) Interest charges on all funds used with respect to the disposal of the completed project, calculated from the date they are likely to be incurred to the date when proceeds from the development are first likely to be received.

Holding costs

(xiv)   There may be costs incurred in holding the site from the date it is acquired until the completion or disposal of the completed development. Such costs may include rates and taxes and insurance costs.

(xv)   Interest charges will be incurred on funds used to pay each item of holding costs from the date it is paid until returns are produced by the completed development, and these interest charges must be allowed for as costs of development.

Miscellaneous costs
(xvi) There may be additional costs involved in such things as the acquisition of the interests of lessees of properties on the development site. The costs of buying out such interests must be estimated, as must the cost of buying out or having modified any restrictive covenants or easements which adversely affect the site for development purposes.

(xvii) There may be a need to spend money on works not actually on the site but nevertheless necessary for development purposes, such as the improvement of street access, the widening of an adjacent road, or the improvement of the mains sewer which serves the site. In some jurisdictions there may also be a requirement to make financial contributions to a local authority in respect of "planning gain" matters, such as payments towards the improvement of car parking in the locality.

(xviii) Interest charges on funds used to pay for the above miscellaneous costs from the date the expenditure will be incurred to the date when proceeds from the completed development are first likely to be received.

Development profit
(ixx) The required development profit will need to be built into the financial calculations, this being an item which may vary over a range of developers all considering the same potential development. Some developers may calculate their required profit as a percentage of the value of the completed development; others as a percentage of the total cost involved in undertaking the project, whilst others may calculate the sum by an arbitrary method to suit their own circumstances and requirements. The above observations on this one item alone serve to indicate the variable nature of the components of the development valuation.

The above summary of typical cost items gives an indication of the large number of items on which financial estimates have to be made when carrying out a development valuation and, because most of them involve the formation of opinions there will almost inevitably be variations in the estimates made by different valuers. It is because of this lack of precision that the courts have generally found difficulty in accepting development valuations calculated by the residual method when they have had the task of determining the value of interests in land in the event of legal dispute on such matters. They have in general taken the view that valuations produced from using such a method provide unreliable evidence as a valuer could, within quite extensive limits, produce the answer he or she wished to submit in evidence by varying some of the many variables until "the right figure" was produced.

In the United Kingdom Lands Tribunal case *First Garden City Limited* v *Letchworth Garden City Corporation* (1966) 200 EG 163 – the Tribunal member said:

> . . . it is a feature of residual valuation that comparatively minor adjustments to the constituent figures can have a major effect on the result

The Lands Tribunal has tended not to accept the method in evidence unless no alternative or simpler method could be used, and the attitude of the Tribunal was clarified in *Clinker & Ash Limited* v *Southern Gas Board* [1967] 203 EG 735, in which case it was said:

> . . . it is a striking and unusual feature of a residual valuation that the validity of a site value arrived at by this method is dependant not so much on the accurate estimation of completed value and development costs, as on the achievement of a right balancing difference between these two. The achievement of this balance calls for delicate judgement, but in open market conditions the fact that the residual method is (on the evidence) the one commonly, or even usually, used for the valuation of development sites, shows that it is potentially a precision valuation instrument.

It is not only in Britain that criticism of the residual method has been made in the courts, and in Australia the Full Court of the Supreme Court considered the method in *Blocksidge* v *State of Queensland* (1991) 2 QdR1, when the three members of the court referred to the limited value of ". . . a notional development method of valuation" which was untested in the market place.

The observations of the Tribunal members and the Court are apposite as the method has to be used with great care if reliable results are to be achieved; results which will enable a potential developer to bid for a development site confident that his or her financial estimates have been made with sufficient precision to ensure that the development of the particular site is likely to give the required return. In reality it is the only truly appropriate method for valuing development sites, and although there are some who maintain that the comparative method is a suitable alternative method, it is suggested that it should in fact only be used to provide an approximate check for results obtained from using the residual method.

There are many who maintain that development sites can be valued using the Method of Comparison, but there are great dangers in using market prices obtained for other development sites as reliable evidence of value on which to base the purchase of a site. A figure paid for another site indicates what another purchaser considered that particular site to be worth at the time of the purchase taking into account the specific features of that site and the needs and circumstances of the purchaser. Planning approvals vary from site to site, physical site characteristics vary, and the quality of the location and environment of each site and the development potential will also be different from site to site, all of these factors having a significant influence on value. The underlying economic conditions will also vary with time as will the supply and demand situation with the various types of property. Additionally, the amount of equity available to each potential purchaser, the rate they have to pay to borrow money, their existing work commitments and other personal or organizational considerations will all affect what each possible purchaser wishes to bid or can afford to pay for any site.

Even when figures paid for other sites are reduced to a "unit of comparison" such as £ or $ per square metre of site area or £ or $ per square metre of developed space permitted on a site, such units of comparison can only be of limited assistance as each is likely to incorporate the type of variables referred to above, which will limit the unit of comparison's direct comparability with another development site.

Although several potential developers may be bidding for the same site, the amount each can afford to bid will vary depending on considerations such as those noted above. To rely, therefore, on the price paid by another purchaser at another point in time for another site, as a reliable basis for one's own bid for a development site would be to place one's faith in another's judgement and to ignore the current economic situation, one's own personal or organizational circumstances, and some, if not all, of the characteristics of the property for which one is bidding. Such action would be foolhardy in the extreme and, even when a comparable is used as a check for a residual valuation figure, it will need very careful adjustment to allow for the factors and possible differences listed above, if it is to be a reliable yardstick for value.

The process of adjustment is itself not easy if a reliable result is to be obtained, and there may be so many variations between the so-called comparable and the subject property that adjustment of comparables may involve even more variables and more subjectivity than the use of the residual method would entail. Additionally, it should never be forgotten that the person or organization which made the comparable purchase might have got it all wrong, ending up with a loss-making project. To use a comparable in such circumstances would simply be to repeat their errors. Accordingly, even though it is difficult to use, it is nevertheless suggested that the only reliable method for individual developers to use in assessing the value of development sites is the Residual Method of Valuation (the Hypothetical Development Method), and examples of the way in which it is used in practice will be considered later in this chapter.

In assessing the *expected market value of a completed development*, evidence of past lettings and sales of similar properties is likely to be the best guide for the valuer. However, care has to be taken to determine whether other transactions related to properties which were indeed similar, that there were no special circumstances surrounding transactions which results in them being unreliable evidence, that trends in prices are investigated as well as actual prices, that demand for such properties still remains, and that there are no circumstances which are likely to cause values to fall between the date of valuation and the date of completion of the development project. In particular, it is important not to assume that what applied in the past will be repeated in the future, even though in some cases it may be. Evidence of past transactions should be used to determine general levels of value, factors which cause variations in value, and trends in the market. Having considered these matters, probably the most important process is to consider today's market carefully to see if it provides better indicators regarding likely future value and, in particular, whether it suggests that changes in the market may be imminent.

In some cases properties may be built to be let out before being sold as investment properties and, if this is the case it will be necessary for both the expected rental value and the appropriate market yield for the developed property to be researched and estimated in order to calculate a Gross Development Value. Comparison with the past letting and sale of other similar properties will be necessary for this purpose and the valuation process would be similar to the following in which the development of four shops is envisaged:

The net rental value of similar shops of 145 square metres net lettable space is £25,000 per annum:

Full net rental value = four shops @ £25,000 pa each =	£100,000
Capitalised @ 8% pa in perpetuity as per market comparison = Years' Purchase in Perpetuity	× _____12.5
Gross Development Value	£1,250,000

Construction costs will depend upon the type of building and the style and design of the specific building or buildings, and will usually be estimated from past experience with similar buildings, from building cost figures obtained from a reliable building cost index, or by obtaining guideline figures from a builder, an architect or a quantity surveyor. Even with such assistance it is still difficult to make estimates which in retrospect prove to have been correct, and variations may result from site or sub-soil conditions proving difficult (and it is frequently almost impossible to accurately determine the precise nature of sub-soil conditions), or from such things as increased labour or building material costs which may occur with the passage of time. Actual building costs are also likely to vary depending upon the type of contract offered to builders, and the more a developer tries to pass some of the construction risks to the builder, the higher will be the price quoted by the builder.

In the examples below the valuer has used a figure in the initial appraisal which is intended to be sufficient to cover all construction work and all site works, and to include such items as site clearance and the provision of roads and services. In later more detailed appraisals the "all-in figure" used here may be broken down into a number of constituent items to enable a more detailed estimate of the likely costs to be made. However, at an early stage in the consideration of a possible project the detailed consideration of the various stages of construction and the various cost items is likely to be impracticable both in terms of the time and the cost required, hence the use of an "all-in figure" at this stage.

Architect's and quantity surveyor's fees are often charged as a percentage of the construction costs, although it is possible for fees to be negotiated on a different basis. For the valuation exercise a reliable estimate has to be made and should either be based on the normal percentage for such types of building, or on the normal rate charged to the developer if he or she has professional consultants who regularly or frequently work for him or her. The same approaches would be used for estimating the charges of any other type of professional consultant who may be required for a development project.

The number and cost of consultants used will vary with the type and scale of the project, but typically the cost of fees will be from 5% of the construction costs at the very low end of the range, to 12.5% of construction costs for more complex schemes which require a high input of time and professional skills. If a developer has "in-house" specialist staff it would be appropriate to make an estimate of the organization's overhead costs (to include a profit element) in providing specialist services.

The *cost of finance* will relate very much to current economic conditions and to the record of the individual developer. The general level of interest rates at any point in time will be determined by international and national economic and financial factors and, within the range of market interest rates at any particular time, the rate charged to an individual developer will depend upon the developer's own past record as both a developer and a borrower. Over and above that, the rate charged is also likely to be dependent upon the overall strength of the development proposal, which will also determine whether funding organisations are in fact prepared to lend money for the specific project. In most cases developers should have a reasonable idea of the rate of interest they are likely to be charged for development loans, although they will not be able to speak with certainty regarding future movements in interest rates. It is changes in interest rates during the development period which often result in the original estimations proving false, and a valuer should carefully consider whether allowance should be made in a valuation for possible future changes. If there is a strong likelihood that such changes will occur, allowance should certainly be incorporated in the initial valuation.

The method used to calculate interest charged on the construction costs in the examples which follow has been used by valuers over many years, and involves averaging the predicted total costs over the entire development period. Not all the required finance will be borrowed on the commencement of a project, and, as indicated earlier, money will only be borrowed from the lender as and when it is required. In practice this generally results in the phasing of expenditure with a lower percentage of total costs being required in the first half of a construction period than in the second half when the major construction work generally occurs and building finishes and fixtures and fittings are paid for. This pattern of expenditure is sometimes referred to as the "S curve", which describes the shape of the expenditure graph indicating an emphasis of major expenditure on site purchase early in a scheme, relatively limited expenditure during the early part of the construction phase, and major expenditure on construction particularly during the second half of the construction phase. In more detailed appraisals at a later stage of a project, estimates might accordingly be made on the basis of low expenditure in the first half of the construction period and higher expenditure in the second half, which might in reality equate to say approximately 40% of the total funds being borrowed for the entire development period.

The "rule of thumb" approach regularly used to make an initial estimate of finance costs takes into account this staggering of expenditure over the development period, and in the examples below the estimated total costs of construction and the fees on construction have been divided by two and the interest charges calculated on the basis that approximately half the required finance will actually be borrowed for the entire development period. While this overall approach can be criticized as being relatively rough-and-ready, it regularly produces reasonably accurate results for an initial valuation. The approach can be refined later in a discounted cash flow format, when attempts can be made to more accurately predict the phasing of the various sums required for construction work and associated fees.

The approach is sometimes varied so that instead of calculating finance costs on the assumption that say 50% of the total funds will be borrowed for the entire construction period, it is assumed that the entire sum is borrowed for half of the construction period, but the writer does not favour this approach as it tends to underestimate the effect of compounding on loan charges.

The fees and charges incurred in leasing and selling a completed development are likely to be based on known levels of charge as solicitors' and real estate agents' fees are regularly based on a percentage of rents or selling prices. If estimates are based on such an approach, the accuracy with which those rents and prices are estimated will in turn affect the accuracy with which fees are estimated. Developers who regularly use the same solicitors and agents may in fact negotiate fees on a different basis such as "so much per house", as used in the examples which follow. If it is contemplated that a property will be let before being sold to an investor, it will be necessary to allow for fees which would be incurred on leasing the property as well as for fees on the eventual sale of the property as an investment.

Estimates of *advertising and promotional costs* should be related to the type of property to be marketed and the degree of ease or difficulty likely to be experienced in achieving satisfactory disposals. Developers, real estate agents, and marketing consultants are all good sources for reasonably reliable estimates of appropriate allowances, particularly with the more common types of property or those with which they have specific expertise. It is important that when economic conditions are difficult the costs for this item should not be underestimated, as advertising and promotion can be extremely expensive, particularly for such properties as major commercial and retail developments.

As implied earlier, the required *developer's profit* is something only the individual developer can determine. It is wise for the valuer to find out exactly what level of profit a developer requires for any project, and then to incorporate that target figure into the development valuation. In instances in which a *contingency allowance* is not incorporated in a development valuation as a separate figure, it must be remembered that the figure for the developer's profit has to be sufficiently large to cover contingencies which may arise, and to ensure that they do not completely erode the profit or make it unacceptably small, for if unexpected costs arise they will almost inevitably reduce the hoped for profit.

If a contingency sum is included as a separate cost item, it must be remembered that this will result in a lower figure being calculated for the value of the development site, as increased costs will result in a lower residual value. While the likelihood of contingency costs arising is very real and should be taken into account, the valuer or developer has to ensure that a situation does not arise in which contingencies are over-provided for, for example by estimating relatively high building costs, by estimating a high developer's profit because of risk, and then by also adding a separate contingency sum. It is possible for contingencies to be allowed for in several places in the appraisal, and the developer must ensure that being over cautious does not result in him or her becoming uncompetitive to the extent that they are regularly outbid for sites in the market-place. There is a need in doing development valuations to balance caution with realism if suitable development sites are to be purchased and acceptable development profits are to be made.

As indicated earlier it may be that there are *miscellaneous costs* involved in undertaking a development, such as the purchase of the leasehold interests of those who occupy existing properties on a site, or the costs of removing or modifying restrictive covenants or easements over a site. In England and Wales in particular, existing lessees may have a right to compensation when properties are acquired for development or redevelopment and, should this be the case, the costs of such compensation must be allowed for as a development cost. Where miscellaneous costs such as these occur, they must be allowed for in the development equation, as must the interest charges which will arise on the funding necessary to cover these costs.

In the examples which follow, costs and values are expressed in £'s Sterling and the costs used were those relevant at the time of writing. The prime objectives of the examples are to illustrate principles and methods, and whilst values and costs may vary with the passage of time, the validity of the principles and methods will not. The same basic principles and methods will likewise be relevant irrespective of what unit of currency is applicable in a specific situation.

Readers may aid their understanding of the principles and methods by converting costs and values to units of their own currency when the £ Sterling is not appropriate, and similarly they may update costs and values as appropriate to allow for the passage of time and changes in prices and values. The writer stresses the fact that the main aim in studying these examples should be to gain understanding of the principles involved and the techniques used.

Example 1

A site is available for the development of twenty detached houses which Developer A believes would sell for £150,000 each on completion. He anticipates that this would be in approximately 18 months time from the present date should he purchase the development site concerned. He does an initial appraisal to show how much he could afford to pay for the site, and he does so on the basis that the cost of finance to him would be 14%. He hopes to make a development profit of 20% of the gross development value.

Value of the development = 20 detached houses
@ £150,000 each =
Gross Development Value   £3,000,000

Cost of development
1. Construction costs – 20 houses of
   160 sq. metres each @ £500 per
   sq. metre   £1,600,000
2. Architect's and quantity surveyor's
   fees @ 10% of construction costs   £160,000
3. Cost of finance on items 1 and 2
   above @ 14% pa for 1.5 years

   $\dfrac{1,760,000}{2} \times 14\% \times 1.5$ years

   1st year – 880,000 × 14% =   £123,200
   2nd year –
   (880,000 + 123,200) × 14%
   for 6 months =   £70,224
   Total interest   £193,424
4. Solicitor's charges and legal
   expenses @ say £1,500 per
   house   £30,000
5. Real estate agent's fees and
   expenses @ say 2% of selling
   price   £60,000
6. Advertising and promotional
   costs @ say £2,000 per house   £40,000

7. Developer's profit at target figure of
   20% of Gross Development Value   <u>£600,000</u>
   Total estimated development costs   <u>£2,683,424</u>
   Balance   £ 316,576

Being the sum available for
   (i)   Purchase of the land
   (ii)   Costs of land purchase
   (iii) Finance costs on land purchase

Multiply by the Present Value of £1 @ 14% for 1.5 years to allow for finance costs	<u>0.819807</u>
Sum available for land purchase and cost of purchase	£259,531
Divide by 1.035 to allow for 3.5% costs of purchase	<u>1.035</u>
Sum available for land purchase	<u>£250,755</u>

Readers who are not familiar with the concepts of the Present Value of £1 as used in the above valuation can find explanations of this valuation process, and others, in A.F. Millington *An Introduction to Property Valuation* (Estates Gazette, London, 1994). It is assumed, in providing these examples, that readers will be familiar with general principles of property valuation and valuation mathematics and methods, but if this is not the case the aforementioned book can also be studied to provide the necessary background understanding. This also applies to the examples in Chapter 14.

If the valuer, in doing the above initial appraisal, had reason to suspect a pattern to be appropriate which indicated that on average only 40% of the funds would be borrowed for the development period, then it would be correct for the estimate in the above valuation to be adjusted to:

1,760,000 × 40% × 14% pa for 1.5 years = £154,739.20

As can be seen, such an adjustment makes a considerable difference to the amount of anticipated interest charges, which would in turn affect the residual sum calculated.

Based on the assumptions made in the above valuation and the results of the valuation, the developer would probably decide to bid £250,000 for the development site. This represents a figure of £12,500 per house block, a figure which is relatively low and which represents only 8.3% of the expected market value of each house. Such a result suggests that the land may only be "marginally ripe for development", or that the developer has been ultra-cautious in valuing the land, in which case he may fail to bid sufficiently high to be able to purchase it.

The costs listed in the above valuation do not comprise a definitive list but are indicative of typical costs which are likely to be incurred in undertaking a development of the type envisaged. The figures used are for example purposes only, and in the real-world they will vary from situation to situation. The objective of a developer or valuer undertaking such an appraisal will be to ensure that all figures used are realistic with respect to the specific circumstances under consideration at the time.

In each development valuation it will be the task of the developer and valuer to determine exactly what work is likely to be involved in the specific project, and then to estimate the cost of completing each specific item of work. As a result, the list of items and the cost of each will almost certainly vary from scheme to scheme. The process of identifying those items has to be undertaken systematically in each development valuation and is a very practical task which requires the valuer to "think through" the entire development project identifying the various items in the process.

It was earlier indicated that this type of valuation has conventionally been used for an initial appraisal and that it could be refined later as more and better information became available. It has also been customary for the figures used in the approach to be based on current information, and so the gross development value is conventionally based on current market values rather than on predictions of market values at the end of the development period. In times of increasing values this is likely to result in low figures being used for the market price of the finished development, but, on the other hand, the approach can also be criticized as it calculates gross development value as if it is receivable today rather than the value being deferred to take account of the fact that it is not receivable until a later date, in this example in 1.5 years time.

For the sake of accuracy, it is suggested that the future development value should be deferred, but probably because in the past there have regularly been increases in values during the period of a development, the over-valuation of one item may have generally been compensated for by the under-valuation of the other. However, in an example such as that above there is little doubt that with most schemes of that nature a developer would aim to sell some of the properties at the earliest possible date, and that rather than all the proceeds being received at one point in time, there would be a phasing of receipts over a period of time. As will be seen later, the use of a discounted cash flow format for the residual valuation enables the phasing of receipts and expenditures to be built into the valuation process in an attempt to provide a more accurate result. However, that can only be done when a reasonable amount of time has been devoted to more accurately estimating both the constituent figures of the appraisal and the dates at which bills will be paid and returns from the development received. In the first instance such precision may not be possible, and an approach such as that used in the residual valuation above is likely to be sufficiently accurate to give an initial indication of the possible success or otherwise of a potential development project.

The figure of £316,576 which is described as the "balance being the sum available for the purchase of the land, costs of land purchase, and finance costs on land purchase" will have to be committed to those three areas of expenditure as soon as the development land is purchased. Additionally, from the date of purchase until the eventual disposal of the completed development, allowance will have to be made for interest charged on the money borrowed to cover those costs, or possibly for the appropriate opportunity cost rate of interest if equity has been used. The device of multiplying the balance by the Present Value of £1 for 1.5 years is a valuation technique for reducing the figure to the appropriate sum which will allow for interest at 14% over the appropriate time period.

A check shows that £259,531 × 14% per annum for 1.5 years is in fact £57,044.91, the total of these two being £316,576, which indicates that the use of the Present Value of £1 has produced the desired division of the figure between the sum available for the purchase of the land and for the payment of interest charged on that capital sum.

The division by 1.035 merely apportions the figure of £259,531 between the actual sum paid for the land and the costs which it has been assumed would be incurred at 3.5% of the land price, namely £8,776.43. Were the costs expected to be say 7.5% of the price paid for the land, then £259,531 would have been divided by 1.075 to get the appropriate split which would have been £241,424 for the land purchase with costs of 7.5% at £18,107.

An alternative method for calculating the division of the sum available for land purchase plus costs of purchase and interest is as follows:

Balance available for		
(i) purchase of land		
(ii) costs of land purchase		
(iii) finance costs on (i) + (ii) =		£316,576

Let the cost of the land =	X	
Then costs of land purchase @ 3.5% =	.035X	
Interest charges = 14% pa on 1.035X for		
1.5 years = .2198* × 1.035X =	.227493X	
		1.262493X

*.2198 is derived as follows:		
Interest in year 1 @ 14% on £1 = 1 × 14% =	.14	
Interest in year 2 = (£1 + .14) × 14% for		
6 months =	.0798	
Total interest on £1 for 1.5 years @ 14% =	.2198	

As 1.262493X = £316,576, therefore X = £250,754.66, giving

Price of land =	£250,755
Costs of land purchase @ 3.5% =	£8,776
Interest charges on the above items =	£57,045

As can be seen this gives the same answer as in Example 1, but this latter method is more suitable if the developer's required profit is based on a percentage of total costs including land purchase and associated costs. If, for example, the target development profit in the above example was based on 20% of all costs, the necessary split would have been achieved by the following approach:

Let the cost of the land =	X	
Costs of land purchase @ 3.5% =	.035X	
Interest charges = 14% pa on 1.035X for		
1.5 years =	.227493X	
Total costs of land acquisition	1.262493X	
Add profit @ 20% on costs =	.2524986X	
		1.5149916X

Were the calculations being done on this basis the divisible sum would in fact be £499,891 and the division would give the following figures.

Price of land =	£329,963
Costs of land purchase @ 3.5% =	£11,549
Interest charges on the above items =	£75,064
Profit of 20% on the above costs =	£83,315
	£499,891

This variation in the method of calculating the developer's profit also illustrates how relatively small variations in approach can result in significant changes to the residual figure.

All the calculations in the above example were done using an annual interest rate of 14% which would be appropriate if interest were charged annually at that rate or if 14% was the annual equivalent rate for interest charged at more frequent intervals. In the modern world it would be unusual in such a borrowing situation for interest not to be charged on a monthly basis or possibly quarterly, and so valuations can more realistically be done on such a basis in a fuller appraisal. However, for the purpose of an initial appraisal, it may still be quickest and easiest for an annual rate to be used, refinement of the approach being made later in a discounted cash flow format valuation when more information is available to the valuer.

As with building costs and other costs, interest rates are likely to change with the passage of time, and readers may find it helpful to test their understanding by adapting examples to include interest rates current at the time of reading.

It is appropriate to compare the results obtained in Example 1 with results which might be obtained by another developer considering the same development scheme and the same development site.

Example 2

Developer B is also interested in the same site considered in Example 1, but he forms a number of different opinions in assessing its value, while his cost of borrowing finance is lower at 13% per annum and his development profit is based on requiring a

25% return on the costs of development excluding the costs of land purchase.

Value of the development = 20 detached houses
@ £155,000 each =
Gross Development Value   £3,100,000

Cost of development
1. Construction costs – 20 houses of
    155 sq. metres each @ £475 per
    sq. metre   £1,472,500
2. Architect's and quantity surveyor's
    fees @ 9% of construction costs   £132,525
3. Cost of finance on items 1 and 2
    above @ 13% pa for 1.5 years

    $\frac{1,605,025}{2} \times 13\% \times 1.5$ years   £163,271
4. Solicitor's charges and legal expenses
    @ say £1,500 per house   £30,000
5. Real estate agent's fees and expenses
    @ say 1.75% of selling price   £54,250
6. Advertising and promotional costs
    @ say £2,500 per house   £50,000
7. Developer's profit at target figure of
    25% of items 1 to 6 above = 25% of
    £1,902,546   £475,636

Total estimated development costs   £2,378,182
Balance   £721,817
Being the sum available for
    (i) Purchase of the land
    (ii) Costs of land purchase
    (iii) Finance costs on land purchase

Let the price of the land =   X
Costs of land purchase @ 3.5% =   .035X
Interest on land price + cost of purchase
@13% pa for 1.5 years = 1.035X x .20345 =   .21057075X
                                          1.24557075X = £721,817

Purchase price of land =   £579,507
Costs of land purchase =   £20,283
Finance costs on above items =   £122,027
Sum available for land purchase   £579,507

The above split has been done on the alternative basis, although the same figures result if the calculations are done using the Present Value of £1 for 1.5 years @ 13%.

Based on the assumptions used by Developer B in his valuation he could afford to bid considerably more for the site than Developer A, and would probably be prepared to bid £580,000. This represents a price for each house block of £29,000 which is over twice the figure A was prepared to pay, representing 18.7% of the estimated market value of each completed dwelling and a more realistic relationship between property value and site value.

A number of different opinions, judgements and facts have resulted in B placing a higher development value on the land. He has taken the view that each house would sell for £5,000 more than A predicted, which results in a higher Gross Development Value being calculated.

He has also estimated that he could build the houses for less money per square metre than A, while basing his building cost estimates on houses 5 square metres smaller than those proposed by A further reduces the estimated building costs. With lower total building costs the figure allowed for architect's and quantity surveyor's fees is lower, but B also believes he could get those services for 9% of building costs rather than the 10% allowed by A, so there is a further reduction in that estimate.

B achieves a considerable saving with his ability to borrow money at a lower rate of interest than that charged to A, resulting in savings on both the construction costs and the land purchase. The fact that building costs are estimated to be lower further reduces the cost of finance.

They both take a similar view of the charges likely to be necessary for solicitors' fees and legal costs, but B estimates that he can get estate agents to work for a lower percentage of the selling prices of the houses, while he also proposes to allow more money for marketing the properties, which may account for his taking the view that higher selling prices will be achieved.

A substantial difference exists in the required development profits, and B is prepared to undertake the scheme for a hoped for profit which is about 25% lower than that required by A, the difference of about £125,000 clearly being a major factor in allowing him to outbid A.

While most of the variations between the two valuations are relatively small in percentage terms, the cumulative effect of a number of different figures in B's valuation (some of which actually involve higher predicted costs) results in him being able to bid 131% more for the land than A's bid. Such examples, which are realistic in practical terms, indicate why courts have been reluctant to accept residual valuations in evidence in litigation, while they also indicate that great care and skill are needed in using the method in the market-place if realistic results are to be obtained. Very good knowledge of building construction and the property development process, good knowledge of property markets, sound judgement, realism, and probably some luck and courage also are required to produce reliable results. Many valuers and developers can in fact get reliable results from the Residual Method of Valuation when used carefully and sensibly, and as observed earlier, it is the only really reliable method for the valuation of development or redevelopment properties.

Sensitivity analysis

As the two examples above have illustrated, the results obtained from development valuations can vary considerably because of relatively small changes in the various component parts of a valuation, and it will be wise for developers to test how reliable the results of valuations are before committing themselves in the market-place to use of the valuation figure produced. They should seek to identify what items are most likely to have an impact on the profitability of a possible development scheme should there be a variation from the predicted figure, and a developer is well advised to test the sensitivity of the projected profit to changes in not only one variable but also to simultaneous changes in a number of variables.

The latter step is wise as market or economic conditions which cause one variable to change, often cause other variables to change at the same time. For example, when interest rates rise the cost of borrowing money will increase, while the rise is also likely to have an adverse effect on the selling price of properties and may result in the cost of construction increasing because of increased costs to builders. It may therefore be essential to test a residual valuation for adverse movements in each of these variables simultaneously to get a true picture of the effect on projected profits of an upward movement in interest rates.

The implementation of sensitivity analysis can be illustrated by reference to the valuation in Example 2 above.

Example 3

Developer B, having valued the development site at £580,000, wishes to test the effect on his projected profit if he bought the site for that figure but the selling price of the completed houses actually turned out to be 5% lower than originally predicted.

Value of the development = 20 detached houses
@ £147,250 each =
Gross Development Value   £2,945,000

Cost of development

1.	Construction costs – as in Example 2	£1,472,500	
2.	Architect's etc. fees – as in Example 2	£132,525	
3.	Cost of finance on items 1 and 2 above – as in Example 2	£163,271	
4.	Solicitor's charges etc – as Example 2	£30,000	
5.	Real estate agent's fees and expenses @ say 1.75% of selling price	£51,538	
6.	Advertising etc. costs – as in Example 2	£50,000	
7.	(a) Price paid for land £580,000 (b) Costs of purchase @ 3.5% of £580,000 £20,300 (c) Interest on (a) + (b) for 1.5 years @ 13% pa £122,131	£722,431	

Total development costs   £2,622,265
Development Profit   £322,735

B's original valuation was based on the assumption of a profit of about £475,636 being made, but it can be seen from the above that a 5% fall in the estimated Gross Development Value results in a projected fall in profit of £152,901 or a fall of about 32%. So the scheme will be highly sensitive to a fall in the selling price of the houses, the percentage fall in the developer's profit being about six times the percentage fall in house prices. The developer could do similar calculations for a 5% rise in house prices and, if it was thought this was the more realistic possibility, then he or she would be well advised to go ahead with a site purchase at £580,000 for the target profit would be likely to be achieved, while any increase in house prices would result in a more than proportionate increase in the actual development profit.

Example 4

Developer B tests his projected profit figure of £475,636 against a possible 5% increase in the cost of borrowing, that is a rise in the rate of interest charged on loan funds from 13% to 13.65% (5% of 13 = .65).

Gross Development Value – as in Example 2			£3,100,000
Cost of Development			
1. Construction costs – as in Example 2		£1,472,500	
2. Architect's etc. fees – as in Example 2		£132,525	
3. Cost of finance on items 1 + 2 above @ 13.65% pa for 1.5 years			
$\dfrac{1,605,025}{2} \times$ 13.65% for 1.5 years		£171,791	
4. Solicitor's charges etc – as Example 2		£30,000	
5. Real Estate Agent's fees etc – as Example 2		£54,250	
6. Advertising etc – as Example 2		£50,000	
7. (a) Price paid for land	£580,000		
(b) Costs of purchase @ 3.5% of £580,000	£20,300		
(c) Interest on (a) + (b) for 1.5 years @ 13.65% pa	£128,504		
		£728,804	
Total development costs		£2,639,870	
Development Profit		£460,130	

B originally aimed for a profit of £475,636, the targeted profit falling to £460,130 after a 5% increase in the rate of interest charged on finance, this being a fall of £15,506 or a 3.26% fall in projected profit. This project therefore has a low sensitivity to changes in the cost of borrowing money as the percentage change in profits is less than the percentage change in the cost of borrowing. This contrasts with the high sensitivity to a change in the selling price of the completed houses.

However, while the project has low sensitivity to an increase in the cost of borrowing money, it should not be overlooked that if interest rates go up it may be symptomatic of a deteriorating economy, and there may be repercussions elsewhere which may adversely affect the project more than the actual increase appears to do. It may well be that the interest rate rise means that purchasers cannot raise so much money, while their loans will cost them more, and this may result in the increase being accompanied by a fall in prices achieved in the market. This may also be accompanied by a longer selling period in a "tighter market", which may mean that the developer has to retain, and pay for, loan funds over a longer period of time. It may well be that these less obvious effects of an increase in the cost of borrowing may be far worse than the apparently relatively innocuous effect of applying the increase in interest rate to the original valuation.

This sort of situation creates the need for changes in a number of variables to be tested, and the availability of suitable computer software enables a large number of changes in variables to be tested relatively quickly. The wise developer will utilize suitable programs to investigate the likely results of changes in both a number of separate variables, and also the likely results of simultaneous changes in a number of variables.

Example 5

Developer B considers what the effect would be on his projected profit if simultaneously there are a 5% increase in the cost of finance, a 5% fall in the selling price of the completed houses, and a 5% increase in construction costs.

Value of development = 20 detached houses @
£147,250 each =
Gross Development Value £2,945,000

Cost of development
1. Construction costs – 20 houses of
   155 sq. metres each @ £498.75
   per sq. metre £1,546,125
2. Architect's and quantity surveyor's
   fees @ 9% of construction costs £139,151
3. Cost of finance on items 1 + 2 above
   @ 13.65% pa for 1.5 years =

$$\frac{1,685,276 \times 13.65\% \text{ for 1.5 years}}{2}$$ £180,380
4. Solicitor's charges – as Example 2 £30,000

5. Real estate agent's fees and expenses
    @ 1.75% of selling price   £51,538
6. Advertising etc costs – as Example 2   £50,000
7. (a) Price paid for land   £580,000
    (b) Costs of purchase @
        3.5% of £580,000   £20,300
    (c) Interest on (a) + (b) for
        1.5 years @ 13.65% pa   £128,504
                                                  £728,804
Total development costs   £2,725,998
Development Profit   £219,002

With adverse movements in three of the variable factors the
projected profit of £475,636 has been reduced to £219,002, a fall of
£256,634, or a 54% fall in profit. This is a substantial percentage fall
in profits, and 5% movements are not particularly large and could
in fact easily occur in a market situation. A developer therefore
needs to be aware of the implication of such movements in the
market which can be indicated by the use of sensitivity analysis. In
the above situation, it is also quite likely that the adverse conditions
may also result in an extended selling period with resultant
increased costs, and there may even be a need to spend more
money on marketing the properties.

Possibly as great a danger exists of a developer's predictions
being 5% out in the first instance, such misjudgements being
relatively easy to make in what is the very difficult task of assessing
development value. Sensitivity analysis is therefore of value in
indicating to a potential developer the implications of
misjudgements in the appraisal process, and overall it is an
extremely useful tool for decision making.

Readers wishing to study further examples of valuations using
the residual method should read *Modern Methods of Valuation* by
William Britton, Keith Davies & Tony Johnson (The Estates Gazette
Limited, London, 1989), or *Valuation: Principles into Practice* edited
by W.H. Rees (The Estates Gazette Limited, 1988) in which Chapter
Thirteen contains numerous examples by John Ratcliffe and Nigel
Rapley.

Shortcomings of the residual valuation approach

The Residual Valuation Method is frequently criticized, and among the criticisms levelled at it are:

(i)   a large number of estimates has to be made by the valuer which increases the chances of error;

(ii) relatively small variations in a number of variables can result in a large cumulative error in the end result;

(iii) the valuation is done at "a point in time" and all estimates by the valuer tend to be made as at that time;

(iv) the failure of valuers to make allowance for changes over the period of the development causes unrealistic results to be obtained from use of the method;

(v)   the method produces misleading results by not using the projected values for rental and capital values as at the expected completion date of the project rather than those pertaining at the start of a scheme;

(vi) the capital value of a proposed development should be deferred to allow for the non-receipt of any benefits from the development until it is completed;

(vii) in using the method valuers do not estimate the costs of construction in a sufficiently analytical way;

(viii) the costs of finance are estimated by a "rule of thumb" method and are consequently calculated in an inaccurate way; and

(ix) the method does not allow for the possibility that income may be received before the completion of a project, for example, if a project is undertaken in separate stages.

Some of these criticisms have been referred to and discussed earlier in this chapter, but there are also arguments which can be put forward in defence of the use of the method. These arguments include:

(a)   the Residual Method of Valuation can be done with considerable speed at a time when a development project is little more than a concept, at which time it is in any event often impossible to value using a large number of variables and great detail;

(b)   its use enables the likely financial success, or otherwise, of a possible scheme to be considered at a very early stage, so helping a developer to decide whether to proceed with further investigations or whether to abandon the scheme without further expense;

(c)   the failure to make what many regard as an adequate allowance for time in the valuation process by predicting future values is balanced by the greater certainty of the present, the use of known figures helping to reduce the inherent risk of development valuations;

(d)   experienced, competent valuers manage to produce remarkably reliable results using the method;

(e)   the basic concept of the method is sound, and as and when more information becomes available to a valuer, so can the use of the method be refined with more variables included and allowance made for time as appropriate; and

(f)   valuers using the discounted cash flow format for development valuations do in fact use the residual method as the concept underlying the calculations remains the same; the major differences being the format in which the valuation is displayed and the inclusion of more variables and more allowances for time in particular.

Some assert that the residual method is "mathematically incorrect", and that it should not therefore be used. The basis for this criticism is not completely clear, although there is evidence that many using the approach in the past have used it in too crude a way and have included mathematical inaccuracies in their implementation of the method, with results which have been flawed for that reason. Poor application of the method cannot and should not be defended, but it is a failing on the part of the user rather than on the part of the method. In practice, the results produced by any valuation method can only be as good as the skill of the user permits. However, the method can be used so that the mathematical parts of the process are accurately done, and the results produced in those circumstances should consequently be mathematically correct.

With respect to the greater accuracy claimed by many to result from the mere use of a discounted cash flow format, it should not be overlooked that criticisms (i) and (ii) above are equally applicable to that approach, while with respect to (vii) and (viii), attempts to be more analytical could in fact result in greater chance of error.

It should not be overlooked that the overall objective of any development valuation is to estimate the likely financial viability of a possible development scheme, and the inclusion of more variables and more allowances for time can just as easily result in a more erroneous result as in a more reliable result. When the discounted cash flow format is used by a valuer, it will only be a more useful decision making tool if the additional inputs used are reliable, and both approaches will only be as reliable as the skill and judgement of the valuer and the reliability of the inputs allow.

Chapter 14

Detailed Appraisal Considerations

It is possible to answer some of the criticisms of the Residual Method of Valuation by adding more variables to a valuation and by making more allowances for time in the calculations. This has always been possible, even in a conventional residual valuation format, but in the past was regularly not done because of the considerable amount of time and cost which would have been required before modern computers and software packages designed specifically for cash flow calculations were developed. The development of computer hardware and software has transformed valuation, as many tasks which were previously extremely arduous have now become relatively straightforward and simple. As indicated in the previous chapter, sensitivity analysis is an extremely useful valuation tool and it is now easy to apply, as a series of calculations which might have taken a day's work as recently as twenty years ago, can now be done in a matter of minutes using a computer.

In a similar way it is now easy to introduce more variables and a large number of different time periods into a valuation using a discounted cash flow (DCF) program, but, when using such a program, the same discounting process and the same interest calculation principles are used as have been used for years in the residual approach. Similarly, when using a DCF format for development valuations, the same fundamental valuation hypothesis is used as that explained in the previous chapter, and the procedures can be varied to find the value of a development site, the size of the expected profit, or the price which a development must fetch on the market if it is to be successful.

The results obtained from use of a DCF approach will, as with any valuation method, only be as reliable as the valuer makes them, and the quality of results will be very much dependent upon the reliability of the information used in the calculations combined with the skill and judgement of the valuer. Unless more reliable results can be obtained using the DCF format than with the residual format, there is no benefit to be gained from its use, so valuers who use the approach have to leave no stone unturned to ensure that the judgements made and the inputs used in the valuation process are extremely reliable.

The ability to easily break the calculations down into shorter time periods is a major benefit of the DCF approach, and it will be remembered that in the examples in the previous chapter using the residual approach, the only times considered were the present and one and a half years from the present, the time difference representing the development period. To seek more refinement through the DCF format, the developer would probably first draw up an intended programme for the development project, which programme might break the eighteen-month development period into segments to allow for:

(i)   drawing up and submitting a development application (a planning application);
(ii)   designing the project and submitting an application for building regulation approval;
(iii)   assigning a date for the acquisition of the development site;
(iv)   assigning dates for the start and completion of site works;
(v)   assigning dates for the commencement of construction and for completion of the various stages of construction, and for stage payments for work done;
(vi)   assigning a date for the completion of the scheme and for its disposal or commissioning for use;
(vii)   indicating the dates at which payments would be made to the various consultants used in the project; and
(viii) indicating the programme for marketing the scheme and for the payment of promotional costs and fees to consultants.

It might be that a valuer would attempt to add even more detail into a programme depending upon the type and size of the project, but even adding the periods and dates which result from the above would provide a reasonably detailed cash flow valuation. It is intended to illustrate the approach by reference to Example 2 in the previous chapter and the residual valuation done by Developer B.

The following discounted cash flow calculations are done on the assumption that the developer intends to bid £580,000 plus 3.5% costs for the land which would be acquired at the end of the fourth month, the first four months being used for obtaining development and building control approval.

Months five and six are assumed to be used for site works with the main construction work starting in the seventh month, a marketing campaign being started in the sixth month with the major emphasis on marketing being in months six to eleven inclusive.

The developer assumes that architect's and quantity surveyor's fees will be paid at the rate of £5,000 per month after larger payments in the first four months for the development and building approvals, and that the first houses will be ready for occupation and sold in the twelfth month, all sales being completed by the eighteenth month.

It is stressed that the following example is a relatively simple illustration of the use of the DCF format to show the development of a residual format into a DCF format and the relationship between the two. A great many more variables could be introduced dependent upon circumstances, and it might sometimes be considered appropriate to use more than one rate of discount, as for instance if both equity and loan funds are used and it is considered appropriate to use different discount rates for each. The overriding consideration in the choice of any variable by a valuer should be that it's inclusion is both realistic and that it helps to produce a more reliable decision making tool.

As indicated in the previous chapter, readers who are not familiar with the DCF format, or whose understanding of it is limited, can find explanation of the approach and of the concept of internal rate of return in A.F. Millington *An Introduction to Property Valuation* (Estates Gazette, London, 1994).

In this example the time periods are shown vertically for ease of reproduction in book form, although in many DCF presentations the time periods are shown horizontally. However, this in no way affects the results obtained from the calculations and, whichever layout may be used it is essential that the valuer includes sufficient description against each entry to enable a user of the valuation to easily understand it.

These calculations suggest to the developer that rather than making the targeted profit of £475,636 a profit of £658,357 will be produced, that is £182,721 more that the potential profit estimated when doing the original residual valuation. The major difference results from the estimations of interest payable on loan funding, the sum estimated in the residual approach being £163,271 interest on construction costs plus approximately £122,131 interest on the sum

End of Month	Particulars	Outflow	Inflow
0	Option to purchase land	£ 5,000	
1	Fees re planning	£10,000	
2	Fees re planning	£10,000	
3	Building approval fees	£20,000	
4	Building approval fees	£20,000	
	Land purchase + costs	£595,300	
5	Site fencing and site works	£30,000	
	Fees	£5,000	
6	Site works	£35,000	
	Fees	£5,000	
	Advertising	£12,000	
7	Construction work	£40,000	
	Fees	£5,000	
	Advertising	£ 8,000	
8	Construction work	£40,000	
	Fees	£5,000	
	Advertising	£5,000	
9	Construction work	£55,000	
	Fees	£5,000	
	Advertising	£5,000	
10	Construction work	£75,000	
	Fees	£5,000	
	Advertising	£5,000	
11	Construction work	£100,000	
	Fees	£5,000	
	Advertising	£5,000	
12	Construction work	£125,000	
	Fees	£5,000	
	Advertising	£2,000	
	Legal & agent's Fees	£8,425	
	Sale of two houses		£310,000
13	Construction work	£150,000	
	Fees	£5,000	
	Legal & agent's Fees	£8,425	
	Advertising	£2,000	
	Sale of two houses		£310,000

Net Flow + or −	Total Debt	Interest Rate	One month's interest
−£5,000	−£5,000	13% pa	£51.18
−£10,000	−£15,000	=	£153.55
−£10,000	−£25,000	1.02368%	£255.92
−£20,000	−£45,000	per	£460.66
		month	
−£615,300	−£660,300		£6,759.36
−£35,000	−£695,300		£7,117.65
−£52,000	−£747,300		£7,649.96
−£53,000	−£800,300		£8,192.51
−£50,000	−£850,300		£8,704.35
−£65,000	−£915,300		£9,369.74
−£85,000	−£1,000,300		£ 10,239.87
−£110,000	−£1,110,300		£11,365.92
+£169,575	−£940,725		£9,630.01
+£144,575	−£796,150		£8,150.03

of Month	Particulars	Outflow	Inflow
14	Construction work	£165,000	
	Fees	£5,000	
	Advertising	£2,000	
	Legal & agent's Fees	£8,425	
	Sale of two houses		£310,000
15	Construction work	£175,000	
	Fees	£5,000	
	Advertising	£2,000	
	Legal & agent's Fees	£8,425	
	Sale of two houses		£310,000
16	Construction work	£200,000	
	Fees	£5,000	
	Advertising	£2,000	
	Legal & agent's Fees	£16,850	
	Sale of four houses		£620,000
17	Construction work	£200,000	
	Fees	£5,000	
	Legal & agent's Fees	£16,850	
	Sale of four houses		£620,000
18	Construction work	£82,500	
	Fees	£7,525	
	Legal & agent's Fees	£16,850	
	Sale of four houses		£620,000
		£2,339,575	£3,100,000
		£102,068	
	Total costs + interest		£2,441,643
	Development Profit		£658,357

used for purchasing the land, the total of £285,402 being £183,334 more than the lower estimate in the DCF format which largely accounts for the substantial difference in the estimated profit.

It is tempting to suggest that such a difference proves the DCF format to be more analytical and more accurate, but this will only be so if the predictions of the dates and sizes of the various payments made in the DCF calculations prove in the event to be

Net Flow + or −	Total Debt	Interest Rate	One month's interest
+£129,575	−£666,575		£6,823.60
+£119,575	−£547,000		£5,599.53
+£396,150	−£150,850		£1,542.22
+£398,150	+£247,300		
+£513,125	+£760,425		
			£102,068.06

correct, while the predictions regarding the selling dates of the various houses are also critical. A substantial part of the interest difference will result from the fact that the residual valuation assumed no returns would be received until the end of the eighteen-month development period, whereas in the DCF calculations the first income from the scheme has been assumed as early as the twelfth month, with regular monthly sales being anticipated thereafter. However, should these predictions be wrong and should income be delayed beyond the predicted dates, the interest charges predicted in the DCF calculations will have been underestimated.

Consequently, while it is true to say that it is easy in the DCF format to allow for more variables and to allow for the receipt of income before a scheme is completed, this will only result in more reliable results if the calculations are based on reliable predictions. It is possible for the addition of more variables and a more analytical approach to time and funding considerations to result in less reliable estimations of value, in that the scope for error could be increased if good judgement is not used or if the additional information added to the valuation is less reliable than the information used in the initial valuation.

In Chapter 4 reference was made to the ***internal rate of return*** as a method of assessing the performance of an investment or a development project. The IRR, as it is frequently called, is the rate of interest at which the net present values of both the positive and negative income flows are equal, and this represents the rate at which an investment earns money. In the above example the IRR is 5.2304% per month or 84.3709% per annum, which would be an extremely good performance from a development. However, the reality might well turn out differently from the predicted scenario and, should the predictions of the valuer regarding the sale dates for houses prove wrong the performance could in fact be substantially different.

If in the above example the pattern of sales was different with one house selling at the end of month 17 and one at the end of month 18, and two houses being sold at the end of each month from month 19 to month 27 inclusive, but all other cash flows remaining unchanged, the IRR would be 2.37489 per month or 32.532155% per annum. This is still a good performance, but it is substantially different from the previous predictions, and further movements against the developer, such as higher costs than expected or an even more protracted selling period, could result in further deterioration in performance, to the extent even of performance being unsatisfactory.

Following an initial appraisal using the residual method of valuation, some developers use a further appraisal stage which seems to fall part-way between the residual and a full DCF appraisal, this being part of the process of developing the residual valuation more fully as they obtain additional information. This appraisal stage involves the costing of various items included under general headings in the residual approach so that many more cost items are actually detailed, the layout of the valuation becoming similar to the following example (without cash sums) for a mixed urban development.

Project income:
 gross office rents =   £
 gross retail rents =   £
 gross residential rents =   £
 gross car park rents =   £
 other rental income =   £____
 Total gross income =   £
Less
 Management expenses =   £____
 Net annual income =   £

Capitalised @ x % per annum (this being × capitalization
the appropriate rate for the development) multiplier____
Gross Development Value =   £

Total Costs of Development
Design and Construction Costs:
 Site clearance and site works £
 Design costs   £
 Cost of construction work   £
 Construction contingency
 allowance   £____   £

Marketing Costs:
 Agency fees and expenses   £
 Legal fees and expenses   £
 Stamp duty etc.   £
 Advertising and promotion
 costs   £____   £

Land Acquisition Costs:
 Freehold or long leasehold
 interest   £
 Purchase of lesser interests
 (eg sitting tenants)   £

Agent's fees and expenses on land purchase	£	
Legal fees and expenses on land purchase	£	
Stamp duty payable on acquisitions	£	
Miscellaneous costs of land acquisition	£ _____	£

Miscellaneous Development Costs:

Rates and taxes due during development period	£	
Purchase of rights of way or rights of light etc.	£	
Costs of buying out restrictive covenants or other encumbrances on the land	£ _____	£

Consultancy Costs:

Project Manager's fees and expenses	£	
Costs of other specialist consultants	£	
Head office administration contribution	£ _____	£

Finance Costs:

On design and donstruction costs	£	
On marketing costs	£	
On land acquisition costs	£	
On miscellaneous development costs	£	
On consultancy cost	£ _____	£

Total Development Costs	£ _____
Anticipated Project Profit	£ _____

Consideration of the above layout shows it to be nothing more than the residual approach with a considerable amount of detail in that separately identified cost items are priced by the developer. However, the only refinement regarding timing which is introduced is the attempt to calculate the various interest charges likely to be incurred to provide separate estimates of finance charges on, for example, design and construction costs and marketing costs. This allows recognition of the fact that there will in reality be different time considerations relevant for different items of expenditure, and therefore varying loan periods to be taken into account. However, the calculations made on such a basis may still involve such approximations as the averaging of construction expenditure over the full construction period.

Such an appraisal format represents an attempt to refine the initial appraisal in the hope of providing a more realistic appraisal on which to base development decisions, and it requires simply one more step to convert such an appraisal into the DCF format.

Wise developers will in fact be refining their appraisals on a regular basis as more and better information becomes available to them, and computer based DCF programs allow this to be done with ease. Such reappraisals should be used by the developer as a guide to changes in approach which may be advisable, and it may well be that important decisions may be necessary even after a decision has been made to acquire a site for development. With reference to the outline appraisal above, it could well be that the site would be purchased on the basis of a mixed office, retail and residential development being appropriate. However, subsequent to the purchase of the site, reappraisal might indicate that, for example, a collapse in retail rents suggests that retail accommodation should be deleted from the development because the cost of creating that space exceeds its newly anticipated value.

Clearly, changes of decision of this nature may not always be possible as it may be difficult to get the development approval changed, or construction may already have progressed too far. Nevertheless, the developer should be constantly reviewing the appraisal on which the initial development decision was based, particularly in the case of developments which take some time to complete, as economic and market conditions may change considerably during the development period. Although fundamental policy decisions may be difficult to change, it could, for example, be relatively easy to note a change in market demand for office accommodation which makes it advisable to upgrade the quality of finish to appeal to a higher quality occupier than that for which the development was originally intended.

Development appraisals should indicate changes in the market to the developer and accordingly should be done on a regular basis, for it is on the basis of an appraisal that the development decision is made in the first instance, and regular reappraisal should indicate both the need to amend policy (if such action is possible), or even to abandon a scheme completely. Although not commonplace, there have been examples in the past of developments being abandoned and demolished when partly completed, to be replaced with alternative developments which changes during the passage of time have made more profitable. The developer in such a situation makes the decision on the basis that the estimated total value of the new development project exceeds the costs of demolishing the existing construction and creating the new scheme plus the current value of the partly completed scheme, with the profit likely to be realized on the replacement scheme also exceeding the likely profit on the original scheme.

Little remains unchanged in the modern world, and although at a past point in time appraisal may have suggested that a development project appeared profitable and the best possible scheme for a site, the development appraisal is just as important in indicating to a developer that change has occurred which may make a different course of action appropriate. For this reason it is important that initial appraisals are refined and regularly updated, for the financial appraisal is the key decision making tool in the development and redevelopment process.

Chapter 15

Project Implementation

Implementation of the project can be considered in five stages, namely the development brief, the design brief, the contract stage, project supervision, and commissioning the project. A developer should in fact form a development programme for each scheme which allocates specific target dates for the completion of each stage, this being an essential part of the overall control of any project. The objective is to ensure that each project is efficiently run, completed, and income-producing by a realistic date. So, for example, in the case of a project which ought to be completed within two years, a developer might draw up the following programme:

(i)   finalizing the development brief – one month;
(ii)   design stage (including obtaining relevant approvals) – five months;
(iii)   offering and letting the construction contract – three months;
(iv)   construction work – fourteen months; and,
(v)   commissioning the completed project – one month.

The precise division of the total time allocated for the development will vary from scheme to scheme depending upon the specific circumstances of each project, but once a programme has been drawn up it will be the objective of the developer to complete all operations within the allocated time periods and, if possible, to complete the overall project ahead of schedule in order to get financial returns at the earliest possible date.

The development brief

The development brief flows naturally from the development concept and it indicates what the developer wishes to achieve. It will be very much influenced by the market research undertaken, which should inform the developer what the market wants, and consequently what should be developed. It will be an extremely important document as:

(a) it will determine the overall objectives of the development exercise; and,
(b) it will define, with a fair degree of precision, what the developer is seeking to achieve with respect to the specific project.

The brief should clearly indicate the objectives in terms of the size and the function of the development, and the uses to be catered for within it. In addition it should clearly indicate the type and quality of location and quality of design and construction to be sought, and the need for general infrastructure provision to serve the development. The approximate size of the site which will be required should be indicated together with an indication of the preferred shape and general features of a suitable site. General design objectives should be defined as well as those relating to the needs of occupiers and users, for example the need to be compatible with neighbouring developments and the need to satisfy broad social objectives as well as specific development objectives. The intended life cycle of the development (with an indication of anticipated major refurbishment dates), the overall quality of the development, and the sector of the market for which it is intended should be clearly indicated, as should the financial criteria on which the project is justified and the financial requirements of the project.

The development brief is the document which indicates in writing to others what the developer envisages and what he or she is seeking to achieve, and accordingly the developer must be the person who determines its contents. It summarises the entrepreneurial objectives as well as the construction aims and, although the developer may be advised by consultants when drawing up the brief, he or she cannot allow the consultants to dictate the contents of the brief as ultimately only the developer can determine whether a brief is acceptable or not. It is the document which the developer can hand to the development team to inform it of the objectives of the project, and to enable the design brief to be developed from it. The development brief should indicate to the team the commercial objectives of the project, it should clearly indicate the developer's objectives and needs, and it should enable the development team to respond more effectively to those objectives and needs.

The design brief

For anything other than small projects the design brief is likely to be developed by a number of members of the development team. As observed above, the development brief indicates relatively concisely the developer's requirements and the objectives of the project, and the design brief will be a further development of that brief, being a substantial document in the case of larger schemes. It should set out in detail the objectives of the development and the criteria to be observed in the design and construction of buildings and ancillary works. The developer must again be the key player in determining the eventual form of the design brief for it will be the developer who finally foots the bill. However, because the document will generally contain a considerable amount of detailed specialist information, the various consultants required by a specific project are likely to be fully consulted by the developer regarding the detailed contents of the brief.

The design brief should contain sufficient detail to enable the architect, quantity surveyors, and other specialist consultants to draw up the detailed plans and costings for a project which will enable the objectives first indicated in the development brief to be achieved, both in terms of the physical characteristics of the project and the financial requirements of the project.

The design brief should provide clear guidelines on the physical specifications of a project, for example the amount of clear working space which is desirable on each level of an office development, the number of floors to be constructed, and the total amount of usable space required. It should clearly indicate cost limits to be observed in design and construction, and should indicate the time frame within which development has to be completed and returns produced. It should give a clear indication of user needs which will have been clarified through market research, so with an office development the quality of lifts, the general quality of finishes, details of any specialist user requirements such as computing facilities, conference rooms, and lighting requirements, should all be clearly indicated if the design team is to ensure that the final design will in fact satisfy market demand. The brief should also indicate the need for car park space and the type of space to be provided, it should indicate the general amenities to be provided in a scheme over and above the major user requirements, and it should specify such details as the requirement for minimum column spacings to ensure that there are sufficiently large areas of unobstructed working space, and also minimum ceiling heights. The last two considerations are likely to be particularly important with industrial and warehouse developments, but are also important in office and retail properties.

Details should be included regarding the quality of fixtures and fittings and of all plant and machinery to be included in the development, for the quality of such items will have a major bearing on the initial costs of development, the costs in use of the developed property, and possibly also the rental and capital values of that property.

It may be that some of the details of the design brief will be dictated by planning and building control requirements, and in particular by special conditions which may be attached to approvals received. Where this is the case such design requirements and constraints should be clearly indicated in the brief.

While the development brief must be sufficiently detailed and precise for the design team to be able to respond effectively to the developer's requirements, it is desirable to leave scope for the architect and other professional consultants to use their own imaginations and professional skills in the actual design stage, otherwise the developer may not get full advantage from the team of consultants. This may even extend to actually suggesting in the brief a number of acceptable approaches to the design process, so indicating to the design team that more than one solution might be acceptable and that their own views and solutions can be incorporated in suggested designs. Ultimately, the developer wishes to get the best development possible to satisfy the development objectives and, having enlisted the services of consultants, presumably because their abilities are respected and valued, those consultants have to be given scope and encouraged to use their own particular expertise in assisting the developer to achieve the development and commercial objectives of the project.

However, whilst seeking to give desirable freedom to the consultants, the developer must retain control of the project and must be prepared to dictate to the consultants, if such a course of action is necessary, to ensure the success of the scheme, particularly success in financial terms.

The construction contract

Once the design stage has been completed, full plans have been drawn up, and the necessary approvals have been received, the developer is in a position to place a contract for the construction work. Because of the heterogeneity of development projects and the varying circumstances and objectives of developers, there are likely to be variations in contractual arrangements from project to project, but nevertheless there are a number of basic approaches to placing construction contracts, each designed to suit a broad set of objectives and requirements.

A developer may offer a *"contract by tender"* to suitable builders or construction engineers. As part of the process of quality control the developer, advised by his development team, or at least the key members of it such as the architect and quantity surveyor, will usually draw up a short-list of suitable firms and offer the building contract to them, inviting each firm to submit its bid for the right to do the construction work for the development.

In drawing up the list of firms to whom the contract will be offered, the developer will be seeking to ensure that no inferior firms are included and that all those invited to bid for the work have the technical ability to complete the construction work to an appropriate level of quality and by the required time. In considering suitable firms, the developer will be guided by the quality of their current staff, and in particular their key staff; their staff relations record and their ability to avoid damaging industrial disputes; their past record in respect of quality and delivering completed projects on time; their past experience in similar types of development to that being undertaken; their past record in projects of a similar scale; and their financial stability and ability to cope with a project of the size under consideration. Their current workload could be an important consideration, as firms which are already fully stretched may find it difficult to pay adequate attention to any further contracts they may win.

The "tender" situation is highly competitive and it will be normal to place the contract with the firm which has submitted the lowest contract figure for the work, obtaining a competitive bid for the work being the major objective of the tender process. In particular this is likely to be the case in a selective tendering situation in which the acceptability of the various firms has been considered previously by the developer, with the result that the only reason any firm is subsequently likely to be considered unacceptable is that their bid figure is too high.

From the developer's viewpoint the advantage of the tender situation is that, unless there is an abundance of work for building firms, with the result that some or all of the firms do not really need work and so do not submit low bids, a highly competitive price for the construction work is likely to be forthcoming. As the control of costs is an essential part of the developer's role, this is a highly desirable state of affairs, but, for the developer to be able to retain tight control of the situation, the initial tender documentation and the contract must be very carefully drawn up with great attention paid to detail. In particular, the responsibilities of the developer, the builder, and others such as the consultants, for each of the various cost items must be very carefully defined, and the responsibility of the various parties for any cost over-runs which may eventuate must also be clearly indicated.

The successful tenderer will have carefully priced the successful bid to take account of the various items detailed in the tender documents, and will be prepared to do only that work for the price quoted, the tenderer's return for work done being the difference between the costs borne by them and the contract sum. Any *"variations"* to the original specifications which are not clearly defined in the contract as being the responsibility of the builder will be charged as extras by the builder to the developer, so it is important for the developer to carefully determine exactly what work is required and to clearly define that work in the documentation prior to agreeing a contract. Making variations to the work by changing design details, changing materials from those originally specified, or adding additional work will result in the builder charging for such work as variations, and such additional items can often raise the cost substantially over the figure first quoted and agreed as the contract sum. It can easily happen that variations may amount to an additional 25% on the original contract price, while the parties to the contract may be placed in a situation of conflict over disputes about variations, which may be damaging to future relationships and the successful completion of the project, and which may possibly result in an extension of the contract period. In addition, increased costs from variations may also result in specialist consultants receiving higher fees if their fees are linked to costs. Careful control of the project will be essential as the placing of a contract by tender almost inevitably creates a situation in which the builder will be looking at every variation to the contract as an opportunity to increase returns for work done, thereby increasing building profits.

With large schemes in particular it is very difficult to include every detail of construction work in the initial documentation. It is also difficult in big schemes to foresee all necessary work at the outset of a project. Variations can occur for reasons outside the developer's control, such as the unavailabilty of building materials or the introduction of new standards by a government or local authority, whilst no matter how careful a developer may be there is always the possibility of omission or mistake in the design and documentation processes. The tender situation is therefore one in which the developer will have to ensure that he or she retains tight control of the development situation if the initial advantage of a competitive bid for the construction work is not to be lost through increased costs for construction variations, additional consultancy fees, and the costs incurred by the developer in ensuring tight management control exists.

It is possible for the developer to place a contract on a *"cost plus contract"* basis under which arrangement the builders agree to do the work for the actual cost of construction work plus a percentage fee payable to them based on those construction costs. In this type of contract many of the observations made about the contract by tender are relevant, as, unless the developer can control the cost of construction work, both that cost and the fee to the builder are likely to increase, while the more the builder can "allow" construction costs to increase the higher will be the return to the builder. Such an arrangement again calls for very careful and very detailed documentation and very tight control of the project by the developer.

In recent years it has become common to get development work done on a *"design and construct"* basis, which has the advantage of normally allowing work to be completed over a much shorter total time period. The same organization is employed to do both the design work and the construction work, as a result of which a development project can generally be completed more quickly because construction planning can begin during the design period, while there is no time lost between the design stage and the construction stage in placing the construction contract. There should also be no problems from difficulties in communication between the designers and the builders as both groups will be part of the same organization so, in theory at least, risk should be reduced in that respect. Moreover, the arrangement should result in there being greater co-operation between the two groups, with the construction group being well placed to actually contribute to the design process. The arrangement also means that the areas of accountability should be clearer, as the number of organizations to which responsibility is delegated by the developer is reduced; the possibility of disputes as to whether the design organization or the construction organization is responsible for problems which may occur in a project is eliminated by one organization being responsible for both areas of work. A major benefit of the arrangement is that the compression of the total development period which should result from a design and construct contract, should also result in the earlier completion of a project with lower finance costs and the earlier receipt of returns from the completed development.

There are disadvantages with such an arrangement, one being that the developer is restricted to the use of the in-house designers of the construction organization with no opportunity to recruit other designers who may have more expertise, more imagination, or more experience with the type of project being undertaken. It is also likely that the in-house designers will be tempted to use a design (or a variation on a theme) which has been used for a previous scheme with the objective of saving time and expense, and also because both they and the construction team are familiar with it, thus reducing the areas of risk for the design and construct team. However, such a solution may not necessarily be ideal for the project under consideration, but the ability of the developer to utilize other designers who might provide more satisfactory solutions will be denied by the type of contract entered into.

In order to reduce the possibility of problems of this nature it will be essential for the design brief to be detailed and unambiguous, and worded in such a way to ensure that the design solutions produced are appropriate for the project rather than perhaps merely being convenient for the design and construction organization. Because the latter organization takes on tasks and risks greater than those in a normal construction contract, the cost of a design and construct contract will normally be higher than that of a conventional construction contract, although the developer will hope that the cost will be less than the combined cost of separate design and construction contracts. However, the main benefit to the developer should result from the time savings with a design and construct contract, and the resultant financial gains. Because two very important stages of the development process are being entrusted to one organization, it is critical that the selection of the design and construct organization is done with the utmost care and that the developer ensures that he or she retains ultimate control of the project.

Whatever the form of contract which is agreed for a development, a developer is likely to try to ensure that the sum agreed for the contract is fixed and that there will be no increase in the construction costs over and above the contract sum. If such an arrangement can be reached, the developer passes to the construction organization many of the risks associated with the construction process, including the risks that labour and materials costs may increase or that costs may increase because of unforseen site or weather problems. It is equally likely that a construction firm will be reluctant to accept such risks and, if it does, it will only do so if the price agreed for the contract is sufficiently high to compensate it for accepting these risks, with the result that for each additional area of risk it accepts it is likely to quote a higher contract figure. From the developer's position there will be a "trade-off" of reduced risk for increased cost, and the eventual details of any *"fixed price contract"* will depend upon how that trade-off is agreed between the parties.

Most developers will also seek to arrange a construction contract which ensures that the builder compensates them for any losses which result from the non-completion of the construction works by the agreed date, and again there will be a trade-off situation on which the parties will negotiate. Non-completion of a project will mean the receipt of returns from the project is delayed, and this may completely destroy the financial viability of a scheme if the delay is too long. It may therefore be that a developer will insist upon *penalty clauses* being included to ensure that if the developer suffers loss because of delay in completing the construction works, or if income flow or sale proceeds are delayed because of late completion, the construction organization will pay the developer compensation for such losses. Clearly, any construction firm is only likely to accept such additional risks if the contract fee is sufficiently high to compensate them for the increased risk, and the developer will again be faced with a situation in which risk reduction will result in increased construction costs.

In contrast to this is the fact that if a project is completed early the developer will be able to receive income or sale proceeds at an earlier date than originally programmed, which will frequently be of great benefit. Accordingly, a contract may include an incentive to the construction firm to complete building work before the date specified in the contract, and a *bonus clause* may make provision for the builder to receive a financial reward, over and above the contract sum, for early completion of the construction work.

While there are *standard contracts* drawn up by a range of professional organizations, such contracts will not necessarily be suitable for all development situations. The final form of a construction contract will very much depend upon the requirements of the particular project, the needs and bargaining power of the developer, and the desire of construction firms to obtain a contract coupled with their own bargaining power *vis à vis* both the developer and other contractors.

Project management

Regular reference has been made to the need for a developer to retain control of a development project at all times, and project managers have been referred to on several occasions. It is common practice for project managers to be appointed by developers as their agents to ensure the efficient implementation of a development project. The project manager will act for the developer in co-ordinating the development team and in supervising and co-ordinating all the stages of the development process, including both the design and construction stages. The role and duties of a project manager were considered more fully in Chapter 5, and in general the project manager will seek to ensure that the developer's interests are best cared for and the development objectives are achieved.

The appointment of a project manager will involve a developer in additional cost, and with small schemes the role of project manager may be fulfilled either by the developer or by the architect, quantity surveyor, or valuer taking on the project management role in addition to their other role in the project. When a project manager is appointed the cost of such an appointment will depend upon the level of accountability which is attached to it. If the project manager is required to take financial responsibilty for some aspects of the project, for instance if the project manager has to take financial responsibily for cost over-runs, then the project management fee is likely to be substantially higher than if the project manager accepts management and supervision responsibilities but no financial accountability over and above those responsibilities. Accordingly, the fees for project management can vary over quite a wide range such as from 2% to 10% of design and construction costs.

In each case the actual fee agreed will depend upon the responsibilities accepted by the project manager and the bargaining power of the parties, that is the developer and the project manager. From the developer's perspective the most efficient project management is likely to be performed by the project manager who accepts some of the risks of the project, as inefficiency will result in loss to the project manager. Although the fees for such an arrangement are likely to be higher, there is evidence to suggest that the possibility of the project manager having to bear losses is a major incentive to ensure efficient project management, as a result of which the savings are generally considerable and far in excess of the additional fees incurred. The use of specialised project managers has, as a result, become normal practice for most modern development projects.

Commissioning the project

If all other stages of the development process have gone satisfactorily, and in particular if there has been efficient project management, this should be a relatively short and simple stage, being the process of taking over the completed construction works and ensuring that they are in the correct state for use by the owner or lessees, or for sale in the market. The works should not be taken over until the project manager and developer are completely satisfied that they are being delivered by the construction firm in accordance with the terms of the construction contract, and careful inspection of the property is important at this stage. It will be regular practice for a construction contract to provide that the developer will not make the final payment under the contract until some time after the building or buildings have been taken over in order to provide time for the work to be tested in use to ensure that it has all been done to the appropriate specifications.

Once accepted from the construction firm it will be the task of the developer or the project manager to ensure that any other work which is necessary, but which was not part of the construction contract, is completed before the development can be occupied, which might include such operations as cleaning, window cleaning, laying carpets, and the installation of various items of furniture or lighting if such are the responsibility of the developer.

Chapter 16

Marketing Developments

The timing of marketing

No matter how good the concept behind a product or how high its quality may be, without appropriate marketing it is unlikely to be as successful in financial terms as could be the case with good marketing. Well-conceived, well-designed and well-made products may in fact be financial failures because of poor marketing, even though they may be offered at competitive prices. As with products in general it is important that property developments are properly marketed, and the marketing programme can often begin almost as soon as the development idea has been conceived.

Indeed, nowadays many development projects are unlikely to proceed past the development concept stage without early marketing of "the concept" and the early acquisition of prospective lessees or prospective purchasers of the completed accommodation. Such a situation arises primarily because many developments are now so large and take so long to complete that no financial institution will provide loan funding unless the likely market success of the concept has been reasonably clearly indicated by the acquisition of lessees and purchasers who have shown an early and firm intention to lease or to purchase the proposed development. Such a situation is also often in the interests of developers who can reduce some of the inherent risk of development by identifying and securing potential lessees and purchasers before commencing the development, then "tailoring" their product to satisfy the identified needs, thereby making the property even more attractive to those interested parties.

Hopefully, well-conducted market research at an early date will not only indicate to a developer whether there is effective demand in the market, but it will also actually indicate those parties most likely to require the use of the space to be developed as well as the most likely investment purchasers once the development is completed and let. Market research should also reveal the type and quality of accommodation required in that user requirements should have been clearly researched if survey questionnaires have been well designed. This being so, the results of that research ought to have been taken into account in the design stage in order to ensure that the developed property fulfils the needs of the market. If such procedures have been followed, unless there have been changes in the market during the development period, successful marketing of the completed development should present few problems.

Marketing tactics

It is important that those most likely to rent or buy the developed property should be made aware of its existence on the market and that they should be provided with sufficient information to persuade them to investigate the product and, hopefully, to purchase it. Depending upon the type of property, various publicity and information methods may be adopted, and usually a combination of methods will be appropriate for marketing. It will be necessary for the developer and the consultants to decide appropriate methods and then to devise a marketing programme, allocating a suitable budget for the purpose.

Typical methods of marketing include display boards on a property; newspaper advertisements; advertisements in specialist journals; radio and television advertising; printed letting and selling particulars; letter-box drops of particulars; postal circulation of particulars to potential customers; radio and television advertising; promotion on computer networks; and promotional events using models, photographs, and video presentations. Clearly the value of the property involved will to a very large extent determine the budget justifiable for marketing, while the type of property will determine the geographical area within which marketing efforts should be concentrated and the type of advertising and promotion likely to be appropriate.

When the major marketing effort is not likely to occur until development is completed, or when development is phased with the result that part of the development will be completed at a relatively early date, it may well be sensible expenditure on the part of a developer to allocate funds to equip a demonstration office or a show-house so enabling potential lessees or purchasers to see exactly what quality of accommodation the finished development will provide. In due course such accommodation can be sold with the benefit of being "fitted out" which may in fact result in partial recoupment of the cost. If early lettings or sales result in part of the development being occupied at an early date, this may well assist the sale or letting of other accommodation, and it may be sound marketing tactics to make early disposals at concessionary figures to create such a situation.

Selling and leasing information

The objective of preparing particulars for potential lessees and potential purchasers should be to provide the information such parties require to assist them in deciding whether the accommodation being marketed is appropriate for their needs and within their price range. If accommodation is not suitable or if it is too expensive, then lettings and sales are unlikely to be achieved, while if marketing particulars are lacking in essential information those receiving them may well be put off pursuing the matter further. The objective should be to assist potential purchasers, and also to put them in a frame of mind in which they are well disposed towards the development and those marketing it.

The location and the type and quality of accommodation being offered is critical information, and location maps and plans of the site and the accommodation on it will be helpful information for interested parties. Other matters of likely importance to a potential lessee or purchaser which should be covered in market particulars include:

(i) dimensions of the site and the accommodation being offered;
(ii) an outline of the construction details;
(iii) an indication of fixtures and fittings and plant and machinery to be included with the property;
(iv) details of the responsibilities of the parties for outgoings and repairs and maintenance;
(v) an indication of the legal interest being offered and of the main lease terms in the case of property to let; and
(vi) information on the expected rents or the asking price in the case of a sale.

It is suggested that a failure to provide such information in marketing particulars is only likely to waste time and money for both the marketers and the potential property users.

A description of the locality in which the property is situated and a pen-picture of the local economy, and details of communications and other facilities in the area are important, as well as photographs or drawings which should give a correct impression of the available accommodation and its location rather than be designed to make them appear better than they really are. The objective should be to provide information which is helpful to those to whom one is seeking to market a property, and misleading information is more likely to antagonize potential lessees or potential purchasers rather than to persuade them to "do a deal". Furthermore, the use of misleading information or the failure to inform a lessee or purchaser of important information might at some future date leave a developer or a marketing organization, or both, open to litigation on the grounds of misrepresentation. Consequently, it is in everyone's best interests that the marketing campaign should concentrate on the provision of accurate, helpful, and relevant information which will assist those on the demand side to decide whether a development satisfies their needs.

The timing of disposals

Reference has already been made to the benefits which can result from agreements to lease or sell reached before a development is begun. Sales or lettings *"off the plan"* have the benefit of limiting the risk to a developer and of ensuring that money returns will be received as soon as a development is completed, the risk of having a completed but empty property which produces no income being avoided. Much of the risk in such arrangements is passed from the developer to the lessee and the purchaser of the development, as each agrees to take a property without being completely certain what the end product will in fact be like. For this reason it is almost inevitable that a sale or letting off the plan will only be achievable at a conservative price, at which figure the purchaser or lessee will be receiving a "discount" for the high-risk taken on by them. In the case of an investment purchaser, a purchase off the plan is only likely to be acceptable if there are also lettings of the accommodation agreed, and the developer's selling price is therefore likely to be reduced both by "rental discounts" provided to lessees and the "sale discount" provided to the purchaser.

A major disadvantage with a sale off the plan is that if prices of accommodation increase during the development period, the developer will be unable to benefit from those price increases having already committed to a sale at a price fixed in the market before the increases occurred.

In recent years it has been necessary to provide a range of *"incentives"* to lessees of many types of accommodation, including rent-free periods of occupation extending to as long as five years, fitting-out allowances to enable lessees to equip accommodation for their own user needs, periods of renting at below the full market rent, contributions to lessees towards the cost of removal from other accommodation, the provision of free car parking spaces, and a range of other incentives. Such incentives have even included developers taking over the responsibility of existing leased accommodation from lessees. In the case of all incentives, it has to be remembered that they either reduce the returns to a developer or they increase the costs of development and, whichever may be the case, their effect is to reduce the potential profit from a development.

It is highly likely that incentives will be an important factor in persuading a lessee to commit to leasing accommodation before it is even finished, so it is important that the financial appraisals should include all costs of "pre-letting" or "pre-selling", including reduced returns and costs of incentives, if an unrealistically high development profit is not to be forecast. The developer ultimately has to decide whether the costs of such arrangements are justifiable or too great, and if they include the need to make too many adaptations to the development or to add too many fixtures and fittings simply to satisfy potential lessees, the costs may in fact become so great as to threaten the financial viability of a development.

The giving of incentives or concessions to lessees in particular may have implications with respect to the tax liabilities of the parties, and this is a matter which should be carefully investigated before such arrangements are taken on board as part of a marketing programme.

In the case of all concessions to lessees in particular, the developer must seriously consider their effect on other lessees and on the potential investment purchaser before agreeing to them, for the capital value of a property may be badly affected by such concessions. Even if a property is to be retained by the developer as an investment this is an important consideration, as the value of the property as a security is likely to be adversely affected also.

Disposal of a development *"before completion of construction work"* is likely to be not quite so disadvantageous to a developer unless there is very much of a "buyer's market" at the time, a buyer's market being one in which there is an excess of available accommodation when compared with potential users in the market for such accommodation. When there is a buyer's market, the developer is likely to be faced with similar disadvantages to those when selling off the plan, but, if on the other hand there is reasonable competition in the market between potential property users, the fact that the property is partly constructed may be helpful to potential users and may help to persuade them that the accommodation suits their needs. There are also the advantages that the developer can be seen to be committed to the project and that the accommodation definitely will be built, whereas with sales off the plan there is no certainty of this happening. In addition, an occupation date for the property can probably be promised with a higher degree of certainty than with a sale off the plan. It is therefore highly likely that a sale at this stage may result in higher returns to a developer than a sale off the plan would produce, particularly if there is reasonable competition in the market and if there are rising prices in the market.

In most market situations the best price obtainable for a completed development is likely to result from a *"disposal following completion of construction work"*. This is because, providing market conditions are not bad at the time and providing the quality of the development is not sub-standard, the design and quality of the finished product will be evident to potential users which should result in them paying a normal market price for it rather than expecting a discounted price to allow for the uncertainty of buying off the plan. The ability to offer immediate occupation is also likely to appeal to many potential users in all but poor property market conditions. While prices obtained following completion of construction work are likely to be highest, particularly when market prices are rising, the downside to the developer of sales following completion of construction is that there is no certainty that all the accommodation will be disposed of; even when property can be disposed of there is no certainty over disposal dates; there is no certainty of disposal prices until a disposal is actually achieved; and there is no guarantee that the market will not "go bad" before disposals can be achieved. In these respects disposals following completion of development represent a high risk situation for a developer.

In most circumstances the decision when to sell will depend upon a range of factors, the most important of which is likely to be the overall supply and demand situation with respect to the type of property which is being developed. Although there have been commentators in the past who have suggested that developers sometimes seek to delay sales to benefit from capital appreciation in rising markets, it is a general rule that a property is not really an asset until it produces income flows or a capital sum from a sale. Consequently, there are likely to be few circumstances in which developers will wish to do anything other than achieve disposals of their property developments at satisfactory prices, and the marketing programme and marketing campaign for any property should be drawn up primarily to take account of market conditions for the type of property concerned and the circumstances and needs of the developer concerned. In most cases one of those needs is likely to be for an early and satisfactory disposal to facilitate the commencement of further development projects.

Chapter 17

Risk and Uncertainty, and Risk Control

There has been repeated reference to the high degree of risk and uncertainty inherent in the property development process and of the need for developers to try to eliminate risk and, when complete elimination is not possible, to try to mitigate the effects of risk.

Development inevitably entails many risks, which include:

(i)   the possibility that anticipated rents cannot be obtained, or even that the property can only be partially let or cannot be let at all, for even though great care may be taken in predicting such matters at the outset of a project, with the passage of time much can change, particularly underlying economic conditions, while unanticipated competition in the market may also arise;

(ii) the possibility that investment yields may increase before a project is completed with the result that there will be a lower capital value than predicted;

(iii) the possibility that a planning authority will not allow as much development as anticipated, or that onerous and costly conditions may be attached to a development or building approval, or that approval may not be received as rapidly as anticipated;

(iv) the possibility that the development site can only be purchased at a higher figure than expected, or that someone else purchases it;

(v)   the possibility that building operations cannot commence as early as hoped for, and also that the building period may last longer than predicted because of such things as bad weather, unforeseen site and construction problems, shortages or delayed delivery of building materials, and industrial disputes in the building industry;

(vi) the possibility that the cost of building labour and building materials will increase, or that new legislation or shortage of supplies may enforce the use of more expensive materials, and even the use of a more expensive design or more expensive method of construction;

(vii) the possibility that the cost of borrowing funds will increase because of such things as increases in the cost of borrowing, a longer than predicted development period, higher than predicted land costs, higher than predicted construction costs, and a longer than expected marketing period;

(viii) the under-estimation of the period needed to dispose of the property with resultant increases in finance costs (referred to above), the need to spend more on marketing, and the delayed receipt of returns from the project.

There are clearly many problems which may confront a developer during the lifetime of a development project, and it will be rare indeed for any project to be completed without any of the above (or other) problems being encountered. A developer decides to proceed with a project on the basis of a series of judgements made about the future, but as the future is inevitably uncertain there is the likelihood that, even with the most skilled and wise developer, some aspects of a scheme will turn out differently from the predictions. A major task for a developer is to try to ensure that as few problems as possible are encountered with any development, and to try to reduce the impact, particularly in financial terms, of any problems that do arise. The developer must try to retain control of the development project at all times, and should implement risk reduction and risk control measures whenever possible. Indeed, a managing director of a large property development company commented that his major role was "the evaluation of risk and the elimination of risk", so emphasising the importance of risk control.

"Risk" is generally regarded as an occurrence or occurrences which can be considered as being distinct possibilities, and the knowledge that certain problems may arise enables precautions to be taken either to prevent their occurrence or to reduce the problems likely to arise should they occur. The possibility that they might occur can generally be foreseen, so, for example, by study of past weather records it is possible to predict that a certain amount of bad weather is likely to occur over a given period of time.

"Uncertainty", as the word implies, refers to a situation where nothing is certain, in which case it is difficult to anticipate happenings and developments, and consequently equally difficult to take steps to counter possible problems which might result from uncertainty. With respect to the example in the previous paragraph, although reasonable predictions could be made of a likely weather pattern over a period of time, the occurrence of the heaviest rainfalls in a century or the longest and most severe drought in a century would probably be impossible to predict. It is therefore unlikely that any steps would be taken, or that any measures could be justified in cost terms, to mitigate the remote possibility of such an occurrence and the disruption to a development project which it might cause.

The developer has to accept that both risk and uncertainty will accompany any development project and, while nothing can be done about uncertainty except perhaps to take out insurance cover with respect to such things as extreme weather conditions and earthquake and fire, all possible steps should be taken to eliminate risk wherever possible, and to mitigate the effects of risk when elimination is not possible.

There is a positive side to the high-risk nature of property development, and that is that it tends to restrict the numbers willing to enter into such a risky business, which in turn restricts competition to the extent that it remains possible to make good profits when developments are done with a high level of expertise. Were property development less risky, competition would probably increase, and potential profitability would accordingly almost certainly be reduced.

There has been regular reference in earlier parts of this book to measures which can be taken to mitigate the harmful effects of risk, and the objective in this chapter is to summarize those measures, to elaborate some of them, and to refer to other measures also. This can be done by considering risks as they relate to a number of aspects or stages of the development process, namely, the developer, timing, the development concept, the development site, planning risks, construction risks, finance risks, and marketing risks. In practice, competent developers will carry out thorough research before commencing a development and will undertake a feasibility study which will cover many of the following considerations.

The developer

As with any activity in life, risk will be reduced if one operates within the range of one's own skills, abilities and experience, and there is always the danger that a developer may encounter problems through undertaking schemes for which he or she has insufficient ability or through simply being over-ambitious.

Such possibilities can be reduced by only developing in areas in which one's expertise has previously been proved, and restricting operations to the type of property which has previously been developed with success, to geographical areas that are familiar and well understood by the developer, and to the scale of development in which success has previously been achieved. Such a policy could well be restrictive and might reduce the possibilities of growth for an organization, so it is natural for developers to wish to expand into other areas of operation and other geographical areas, but when doing so caution should be exercised and extremely thorough research undertaken at the outset.

The possibility of risks being encountered through the actions of the developer can be reduced by the use of appropriate specialist consultants of the highest quality, very careful choice of the development team members, the choice of a high quality project manager when one is used, and the employment of a highly competent and reliable builder with successful experience on the type of project being undertaken.

The possibility of problems resulting from being over-stretched financially can be reduced by ensuring that the financial gearing of individual projects is such that, even if problems are encountered their financial viability is not destroyed, while the number of separate projects in hand at any one time should be such that the capital resources and the personnel of the organization are also not over-stretched.

Timing

As observed earlier in this book, extremely good property developments can be financial failures simply because they arrive on the market at an inopportune time when demand is low or even non-existent. A downturn in an economy, or even simply a downturn in demand for a particular type of property, can result in a property being offerred on the market at a time when it is impossible to achieve realistic rents or a capital figure which covers the cost of development. This can occur despite the fact that, at the time the development was begun, there was strong demand for such properties, and it is critical that a developer should ensure that all rental and capital value projections are based on the anticipated market conditions at the time the property is likely to be completed, rather than being based on market conditions at its initiation. Such projections must be realistic if a development valuation is to fulfil its objective of indicating to a developer whether a project is likely to be financially successful.

Careful research must be undertaken with particular attention being paid to underlying economic conditions and in particular to economic trends, while the same should be done with respect to the sector of the property market in which development is proposed, both the type of property and the locality being carefully researched. In this respect the proposals of other developers are important, for current shortages can soon disappear if a number of developers seek to satisfy the same market demand, while if deteriorating economic conditions simultaneously cause reduced demand, surpluses of supply can soon result.

A developer should be seeking to supply property onto the market at a time when economic conditions are favourable and demand is increasing, rather than at a time when economic conditions are deteriorating and demand decreasing. Judging this accurately is not easy, and it is particularly difficult if a large development is undertaken which takes a number of years to complete. However, good timing of product delivery is critical to financial success and any miscalculation can result in an otherwise excellent scheme failing financially. There were many examples in the late 1980s and early 1990s in many countries of extremely well conceived projects, which were also well designed and well constructed, being financial disasters because they were completed at a time of economic recession or depression.

Where it is possible – for example with estates of houses or warehouses or factory units – some of the risks relating to timing may be reduced by completing a scheme in stages, and offering each stage on the market as and when it is ready. Any reduction in demand or any shortfall in market prices should be detected at a relatively early stage if such an approach is adopted, so enabling a developer to alter policy to take account of unanticipated market conditions. Obviously, it is difficult, if not impossible, to adopt such an approach with high-rise office or residential towers and other similar developments.

The development concept

the development concept is ill conceived it will be extremely :ky if a project turns out to be a financial success. The developer eds to determine through careful market research whether there effective demand for a particular type of property, and if so to get accurate indication of the user needs.

A proposal to develop a new shopping centre may, in the first instance, appear sensible, but if research indicates that although there is currently unsatisfied demand it is in fact only limited, that other developers are proposing new developments, and that the local population structure is changing with falling levels of income and altered consumer preferences, then it may be that the concept is in reality ill conceived and likely to fail. It is therefore important that thorough market research should be undertaken by competent staff or competent consultants to avoid the possibility of embarking upon an ill conceived project.

When effective demand and need are both indicated it is important that the concept is strengthened through careful choice of location and careful site selection, and that the project is aimed at the appropriate market sector, that it is of an appropriate size to increase the prospect of success, that it is well designed, and that it is offered on the market at an affordable price. The preparation of good development and design briefs can help to achieve these objectives. Again, risk can be reduced by the use of good staff and consultants.

The development site

No matter how much unsatisfied demand there may be in the market for space of a specific type, and no matter how good the development concept may be in other respects, the selection of an unsuitable development site can threaten the financial viability of a project. This threat may result from a secondary or poor location depressing the market value of the completed property, or it may also result from unreasonably high development costs resulting from difficult site conditions. The latter might, for example, occur if there is poor drainage, poor sub-soil conditions, and restricted load bearing qualities which result in foundation work being too expensive. A poorly shaped site, or a site which is too small, may result in the building design being inefficient and costly in relation to gross development value.

To avoid such problems the developer must again undertake careful research to ensure that the type of property in demand is clearly identified, and that the most appropriate sites are considered for the project. With respect to physical site characteristics, careful site tests should be undertaken before purchase to reduce the possibility that the physical nature of the site may prove to be unsuitable. A carefully worked financial appraisal should indicate the maximum figure which a developer ought to bid for a specific site, and that figure should not be exceeded by the developer.

It should be remembered that getting an inferior site at a low purchase figure will not improve the quality of the site, and if no site of appropriate quality is available on the market then the development concept should be abandoned.

Planning risks

The major risks that are likely to arise at the planning stage are that insufficient development will be allowed on a site, that onerous and costly conditions may be attached to an approval, that expensive planning contributions may be required by the approving authority, or that no approval at all may be given. There are also the risks that planning procedures may be lengthy and costly, and that building control design requirements may result in extra cost.

To counter such risks no site should be purchased until satisfactory planning and building approvals have been received and, before completing a purchase, a careful development valuation should be undertaken, based on the details of the approval received, to ensure that profitable development can be undertaken with such an approval.

The risk of costly but abortive work being undertaken can be reduced by early discussions with the relevant control authorities, while the risk of getting satisfactory approvals, but of not being able to secure the development site at an appropriate purchase price, can be avoided by use of an option to purchase or a conditional contract for site purchase prior to doing all the necessary work for approval purposes.

Construction risks

The main risks with respect to construction work are that site conditions may result in unanticipated site works being required, the start of construction may be delayed, completion of construction may be delayed, construction work may cost more than estimated (possibly because of increased labour and materials charges), and that other problems such as the non-availability of materials, industrial problems, and bad weather may result in an extended construction period with resultant cost increases and delayed receipt of returns from the completed development.

Efforts can be made to avoid site problems by thorough site investigation prior to purchase and, as discussed in Chapter 15, efforts can be made to avoid all the other problems by careful selection of the building contractor, by use of a high-quality project manager and development team, and by use of contract terms designed to pass responsibility and accountability for these areas of risk, and resultant price increases, to the building contractor.

Finance risks

The main areas of risk with respect to finance are that loan funds will not be available or will be insufficient, that the rate of interest payable on loan funds will be too high, that that rate will increase during the development period, and that use of loan funds will be withdrawn before the completed development is successfully marketed.

Risk reduction measures include ensuring that sufficient funds will be made available on suitable loan terms before any significant commitment in financial terms is made by the developer to a project. The developer should seek to ensure that the funds will be sufficient to allow successful completion of the project with a provision for contingencies such as cost over-runs or extended construction or marketing periods. The developer should seek loan terms which are not unduly onerous and which ensure that the rate of interest charged for funds will not be increased, while the loan period should be sufficient to allow for the predicted development period plus a contingency period to allow for delays in completing construction or difficulties in disposing of the development.

Marketing risks

The major risks likely to be encountered in marketing a development are that rents achieved for accommodation will be lower than predicted, that the gross development value will be lower than predicted, that the development cannot be disposed of without delay following completion of construction work, or that either some or even all of the accommodation cannot in fact be successfully marketed.

Careful and early market research to clearly identify the market sector to be targeted, and the likely needs of that sector at the date of completion, is an essential risk reduction measure. Early marketing can seek to identify potential lessees and purchasers, and efforts can be made to ensure that they commit themselves to leasing and purchasing before the developer makes substantial financial commitments to the project. These matters are considered more fully in Chapter 16.

The costs of risk reduction

All of the risk reduction measures considered above have cost implications to a developer, and increased costs inevitably have an adverse implication for the potential level of profit, while some of the measures will also reduce the market value of the completed development. The greater the number of consultants employed and the more competent they are, the greater will be the sum required for consultancy services. The better the site purchased, the more it is likely to cost, but that additional cost is likely to be countered by the production of a more valuable finished product, and possibly by cheaper construction costs also.

Insistence upon contracts for construction work and for borrowing funds which pass some of the risks to the building contractor and the financial institution will almost inevitably result in their charges being increased, which again will reduce the size of the potential profit, while committing to early lettings and sales will almost certainly result in those having to be at favourable figures for the lessees and purchasers, which will once more reduce the potential profit from a development.

It should not be forgotten that all the additional costs resulting from these measures should be countered by the reduction of total risk attaching to a project, and they should also be compensated for by some increases in the value of the completed development which result from such factors as better site selection and a product which is specially designed to satisfy market demand. There is a "trade-off" situation for the developer in which a decision has to be made as to whether the benefits resulting from the measures taken exceed the costs incurred. The main decision-making tool for the developer in this process will be a series of development appraisals based on a range of possibilities and, should there be an acceptable or most favoured possibility, it should be carefully tested with sensitivity analysis.

Although every possible effort has to be made to reduce and control risk, at the end of the day developers who spend too much on risk control and are too conservative in an effort to reduce risk may become uncompetitive, as a result of which they may be unable to bid high enough to purchase any development sites, being consistently outbid by other less cautious developers. It should also be remembered that it is market competition which determines the market values of developed properties and, even if developers succeed in purchasing sites, they cannot necessarily pass on to property lessees and purchasers the costs of controlling risk, which costs, if excessive, will reduce the development profit to the extent that it may in some cases become inadequate. Developers therefore have to use their judgements to determine an acceptable balance between the costs and benefits of risk control measures.

All of the above considerations will almost inevitably be incorporated in a feasibility study made by the developer, details of feasibility studies being considered more fully in the following chapter.

Chapter 18

Feasibility Studies

Reference has regularly been made to the need to conduct careful research and for all aspects of the development process to be thoroughly researched before a commitment is made to develop, and prior to any significant financial expenditure being made by the developer. Such research will generally be incorporated into what is regularly referred to as a *feasibility study (or viability study)*, and it is the intention in this chapter to consider the various stages which might be incorporated in such a study for a major development project. In doing this many of the matters considered earlier in this book will be recalled in summary form.

The Concise Oxford Dictionary definition of "feasible" is "practicable, possible; (colloq) manageable, convenient, serviceable, plausible". The objective of a feasibility study for property development purposes is quite simply to determine whether a particular type of property is needed in a specific location, and also whether the development of such a property is likely to produce a profit in financial terms. Will the project be practicable, possible, manageable, and plausible in financial terms?

The depth of any feasibility study will depend to a very large extent upon the size of the proposed project and the amount of money likely to be involved in the development. For small projects with limited budgets an extensive study will not be justified, but it will nevertheless be important for a feasibility study to be undertaken and for it to concentrate on the most important factors in the specific case. Although a project may be of limited value, it is still essential to determine that it is practicable in physical and financial terms, and the limited size of a scheme will in all probability mean that essential research is easier to undertake and consequently less expensive.

Research needs to be searching and should be undertaken into all matters which are likely to, or which might, affect the profitability of a development. With respect to the amount of information collected in the research process, this again will vary dependent upon the size of the project and, although the collection of too much information is likely to increase costs, it is better to err on the side of being too well informed rather than being under informed. Information which is not immediately usable may well be useful at some future date, and the failure to collect sufficient information may result in critical considerations being overlooked.

The underlying economic scene

Economic conditions are critical as the demand for property and property values invariably increase with economic booms, while economic downturns, recessions and depressions invariably result in reduced demand for property and reduced property values. Investigating economic indicators and assessing what future trends in the economy are likely to be should therefore be the starting point in any feasibility study for property development purposes.

It is important to consider the *international scene* in terms of both economics and politics, as with the extent of international trade in the modern world the economies of different countries are to a very large extent interdependent and, what happens, for example, to the American dollar or the Japanese Yen, or to interest rates, employment levels or production in those two countries in particular, can have profound effects on the economies of a large number of other countries.

If a country is heavily dependent upon the export of primary products, current international prices and trends in those prices can be important indicators for the home economy, and there are probably no countries for which international factors are not important determinants of home policy and the well-being of the local economy. Any indications of political instability in other countries, and the possibility of strife between countries, could be important considerations for a prospective developer trying to predict the future of a local economy. International considerations are particularly important in view of the large number of large companies which are now international in nature with the ability to switch investment and production from one country to another at fairly short notice, with consequent implications for others who may be dependent upon such companies for their own financial success. Being international in nature also makes many companies vulnerable to adverse conditions in countries other than that which may be under current consideration.

Any factors which are likely to affect the decision to place deposit funds in one country or another may well affect both the availability of loan funds and their cost, and such considerations are particularly important for most property developers. International exchange rate figures may be important if the possibility of using international finance is under consideration, and also if the possibility of carrying out development abroad is being contemplated. Movements in exchange rates may also be important as they may affect both the values of overseas assets and the relative values of assets held at home and abroad.

The *national economic situation* is extremely important for property developers who need to consider both what is the current state of the economy and what are future expectations. Important economic indicators to be researched are Gross Domestic Product (GDP) figures and trends in particular, and interest rate levels and trends, for they indicate what the current state of affairs is and they affect both the cost of development finance and the ability of the public to spend. The rate of inflation, the levels of employment and unemployment, and trends are important, as they affect the ability of consumers to spend and also the level of economic activity overall. The level of savings is important as it is an indicator of general economic prosperity and of the ability of purchasers to invest in property developments.

The level of foreign investment in a country is important as a high level indicates that foreigners have confidence in a country and presumably in its economy, while it also provides another source of property investors if local laws permit investment by foreigners. However, it also needs to be considered whether the source of foreign investment is a stable or unstable source, and whether foreign investment funds may in fact only be temporarily lodged in a country with the possibility or likelihood of such funds being withdrawn at short notice.

The level of international debt may be an important indicator as it may greatly influence future trends in an economy, but high international debt is not necessarily a negative factor as it is normally the rule in lending that funds will only be lent to those who are generally regarded as stable and reliable. High international debt may therefore be an expression of confidence by other countries, but, if such debt becomes too large, a national economy may be adversely affected by the need to devote too much productivity to funding that debt with adverse effects on domestic activities.

The economic competitiveness of a country *vis à vis* other countries is important as the possession of good natural assets or a well-developed and competitive industrial sector or technology sector may give fundamental strength to a country's economy. Overall, in looking at both international and national economic factors, the researcher is seeking to determine the level of strength in the national economy, and the likely future performance of property developments in the light of the overall economic scene. It is extremely likely that not all indicators will be as positive as desired, but that does not mean that development or investment are not advisable. For instance, even in times of high interest rates or high unemployment, development and investment may be advisable if the diagnosis is that interest rates are likely to fall and employment to rise, with consequent future improvements in the national economy being a reasonable expectation

It is the writer's view that many past decisions to undertake major property developments have been made without adequate consideration of national and international economic indicators which, if taken into account, might have resulted in many development failures being avoided. The importance of such indicators should not be ignored or underestimated.

With respect to *local economic considerations*, many of the considerations referred to under the national economy will also be important at the local level. In addition, the type of local economy which exists will be important, for example is it primarily agricultural, industrial, commercial, or a mixed economy? If it is highly specialized, are there any signs of a wider range of activities being developed to reduce the dependency of the local economy on one or a limited number of specialisms which may make it particularly vulnerable to possible economic downturns in those specialisms?

The current state of the local economy when compared with the national economy should be considered, as, even during the recessionary conditions of the early 1990s in many countries, some areas were relatively well shielded from the effects of downturns by virtue of the nature of their local economies. An example was Canberra in Australia which, as the seat of the Federal Government, has an extremely high proportion of public-sector employees, with the result that even when unemployment rates elsewhere in Australia were in double figures, the rate in Canberra remained in low single figures making the local economy more stable during the recession. However, "on the other side of the coin", should a government decide to reduce the level of government employment and perhaps to privatize many activities previously undertaken by the public sector, or to relocate some of those activities, a local economy such as that of Canberra might become vulnerable to such changes.

Statistics of employment categories and the characteristics of employment would be helpful to a developer, the dependence of families on two incomes and the dependence on a single-product economy or even a single local employer being the type of information which could be of great importance in making a decision whether to develop in a locality.

Property market study

This will be an extremely important part of a feasibility study and assessment of the *current stock* is probably the best starting point. It will be necessary to determine whether there is a shortage of supply of a specific type of property, a balance between supply and demand, or a surplus of property. The absorption rate on an annual (or other periodic) basis should be determined, and an assessment should be made of how long the current state of the market is likely to continue and when it is likely to change if change appears probable.

The *real market rent levels* should be determined which should take into account not only the rents passing but also any concessions or inducements which have been given to tenants in the past. In making this assessment as much evidence as possible should be obtained, and it should be closely examined to ensure that all the facts relating to each letting are known. Suspect evidence should not be used. Analysis of market evidence should hopefully indicate whether rental growth is occurring or is likely to occur in the future, and the rate of increase, or decrease if rents are falling, should be revealed.

The *current capital values* of the property type under consideration should be researched from actual market evidence wherever it exists and, as with rents, every effort should be made to ensure that all the relevant facts relating to sales and purchases are discovered. The quality of the available evidence must be closely investigated to find out whether there were special factors affecting transactions such as the existence of marriage value with adjoining properties, the merging of legal interests in a property, sale to a subsidiary company, or other similar circumstances. Where there were special factors it will be necessary either to adjust the evidence or to discard it as unrepresentative and unreliable. ("Marriage value" is an increase in value over and above the separate values of two or more interests in land which results from uniting the ownership of the various interests in circumstances in which the value of the united interests is greater than the total value of the separate interests.) The objective will be to find reliable market yields for the type of property, or alternatively reliable capital values per square metre, which can be applied to estimate the development value of a new property. In using evidence it has to be remembered that recent market evidence is likely to be the most relevant, whilst it is the value at the likely completion date of a development which has to be estimated.

Land availability and evidence of land values for the type of development being contemplated should be investigated, with the location, the size of available blocks or plots, the physical features of available land, the availability of services to each block, and the availability of suitable access being important information. In assessing the value of available land, any comparable evidence should be considered for guidance, as should variations between comparable sites.

The typical *costs of development* for the type of property should be researched, taking into account locational factors, the state of the local construction industry, current workloads of building contractors, topographical and geological factors, and trends in the prices of labour and building materials.

Market *workspace ratios* should be investigated taking into account both statutory requirements and local market practices, as while statutory requirements might say that ten square metres per office worker is a minimum space requirement, the type of user in a locality might be such that a far greater allowance per worker might be common practice. Trends are also important and if there is a trend towards space sharing between office workers this might result in future falls in average space allocation per employee.

Proposed developments should be researched, as any developments of a similar type which are currently approved or which may be proposed may seriously threaten the viability of a particular development proposal. Depending on the type of use, it may be that proposals outside the immediate locality should be taken into account, as the development of a major new retail complex twenty miles away in another town may steal customers from a locality. This could also be relevant with a major office or industrial developent, and the ability of the market to absorb all likely developments of a particular type of property must be fully considered. The scale and type of each planned development and the market sector at which it will be aimed should be determined.

The size of the *catchment area* should be considered both in terms of geographical limits and communications networks and travel considerations. Where travel between areas is particularly easy, competing centres of population may exert a strong attraction to the population of another area, and the proximity of settlements to each other and the ease and cost of travel between them should be investigated. Such considerations are particularly important with retail developments and, as well as the danger of losing trade to another area, the possibility of being able to attract custom from other areas should not be overlooked. Proposed improvements in road and rail links and growth in population may also result in a catchment population increasing in market terms.

With respect to retail developments, *retail expenditure data* should be obtained and studied to determine the total expenditure in an area, the type of expenditure, trends in expenditure, and details of expenditure escaping to other areas together with reasons for that happening. The possibility of attracting such escape expenditure should be considered.

The overall objective of the property market study will be to identify any *areas of under-provision* which should also generally be *areas of opportunity*. Such will often arise where there are pockets of population for which there is currently under-provision, where the size of population is increasing, where the affluence of the population is increasing, where there is currently under-provision of a specific type of development, or where there is a need for improving the existing style or class of property provided. Again, when such matters are being considered trends and changing consumer preferences are very important and should be noted by the researcher.

Demographic study

This will be very important for almost all types of development and should establish the current population size; trends in population and any likely changes in current trends; the annual growth rate if the population is growing and the rate of decrease if decreasing; the reasons for changes, for example whether growth is caused by birth, by immigration or by special factors such as the establishment of new towns; the density of population and the ability of settled areas to absorb more population; the geographic distribution of population; the age structure of the population and details of any changes in population structure; distribution of population by sex and by marital status; and details of family size, the numbers, sexes, and ages of children being researched.

Depending upon the type of development being contemplated, some of the above demographic details may be more important than others, and it will be the task of the researcher to ensure that at least the most important demographic information is available and that it has been updated to provide sufficiently reliable information about the present, and a reliable basis for predicting the future.

For retail development in particular, employment data is also important and should include the type of employment (for example "blue collar" or "white collar"); details of male and female employment; the levels of earnings and incomes; the levels of disposable income; current spending details to include the levels of expenditure, the type of expenditure, the types of goods purchased, and geographical expenditure patterns; and the predominant features of demand. Some of this information will also be important in indicating the availability of labour for office and industrial development.

As indicated in the section on the property market study, the overall objective is to try to identify where there are gaps in the market which will justify undertaking new development.

Financial factors

The feasibility study should provide information which enables the level of returns from property development and property investment to be adequately considered and to be compared with returns available from other investments. Returns from equities and the condition of equity markets should be investigated, as should government bonds, local authority bonds, and interest rates available on bank deposits.

The volatility of each type of investment should be considered, as should market trends. In general terms, if yields are low they are generally likely to rise in the future, while the converse also applies.

Potential tenants and property rental levels

Rental levels were referred to under the property market study section earlier in this chapter, and if a development is likely to proceed, there will be a need to try to identify likely tenants, to research their specific needs, to investigate their ability to pay specific levels of rent, and, more importantly, to investigate their willingness to pay those levels of rent.

The levels of rents passing on existing properties are useful indicators, but it is unfulfilled demand which is critical with new development, so those in the market who currently require property should be researched whenever possible, particularly as it is the level of future rents which the survey is trying to identify. It is also trying to identify the type of unfulfilled demand, so, if there is a shortage identified of office accommodation, it will be important to identify whether the demand exists for small suites of accommodation or for major areas of space, or for luxury accommodation, middle-market users, or other types of user. The latter is important as, with all types of property, the provision of better quality space will not necessarily result in higher rents being paid.

With respect to the development of shopping centres or other large areas of retail space, it is advisable at an early stage in the research to try to identify and, if possible, to secure major tenants as lessees. The future success of a centre is likely to depend upon the existence of lessees who are "magnets" so helping to attract customers to a centre, and it is important to try to secure a number of important lessees of this type. "Anchor tenants" – that is tenants who are substantial and who will provide a solid basis for the economic success of a centre – should be identified for key positions, and efforts should be made to ensure that a sufficient number of lessees are identified and secured before a commitment is made to proceed with a major retail development. It is also important to ensure that "pre-commitments" from lessees are likely to result in a good retail mix being established for a centre to enhance its likely trading success.

It is desirable to try to establish through such research the likely market needs of the different retail outlets, such as supermarkets, food shops, market shops, fashion outlets, and other specialty trades, as well as the needs of service tenants such as banks and travel agents. Such information is invaluable at an early stage as it has important design implications with respect to the number of shops to be provided and their sizes and distribution throughout a centre. For this reason the need to establish the seriousness of the interest shown by potential lessees is very important and, if possible, they should be asked to commit themselves to leases in the property to be developed.

This is also an important consideration in the development of a major office complex, and it is desirable to identify a major tenant for a large area of the total space to be developed and to obtain their pre-commitment at an early date. As with retail property, the type of likely lessees and their specific needs should be researched, and the office space requirement per employee should also be determined, for such facts will again have important design implications.

With major developments all market evidence should be carefully researched, and researchers should be wary of hearsay evidence, as there may be unknown facts or special factors which relate to such evidence but which may be unknown to the researchers. The feasibility study is an extremely important part of the risk control process; identifying market interest and the type of accommodation required, identifying specific lessees, and securing their commitment to take space, are major steps in risk control.

Identification of investment purchasers

This is another important risk control step as many developers develop to sell and take a capital profit on completion of a development. Indeed, many may not be able to afford to retain developments even if they wished so to do, but others see their occupation as being to develop rather than to invest, and they also see greater total profitability in redeploying their equity to make a series of capital profits from different development schemes. The existence of "end-purchasers" for developments is therefore very important to the financial success of many developments, and early identification of potential investment purchasers through market research is an important step in the overall development process.

Having determined the type of property demanded by potential occupiers, the researcher should seek to identify investors who might buy such a property occupied by lessees of that type. Likely investors should be identified by name, and whenever possible the size of their holdings and details of their present portfolios should be considered to see whether the property to be developed might fit into their investment needs and investment preferences.

Likely interested investors should then be approached to investigate their current need for property investments, the type of investments needed, and their ability and willingness to purchase. The latter is likely to be affected by the size of the scheme and its location, the quality of the scheme, the existence of and quality of pre-committed lessees, and their ability as potential investment purchasers to influence the actual design and management of the development project. If they are to pre-commit themselves as investors they are likely to want, if possible, to get a "tailor-made investment" at an appropriate price in return for their commitment to buy, and this is likely to require their participation in the design process.

When the most likely investment purchaser is a foreign investor their level of commitment and the ability to hold them to a commitment should be considered. If their initial interest results from an inability to get attractive investment returns in their own country coupled with higher returns on the proposed development and favourable exchange rates, it should not be overlooked that such factors could change over a short period and adversely affect their level of commitment.

Development approvals

The feasibility study should carefully investigate the planning situation as a scheme will need to be acceptable to both the relevant public authorities and the public in general. Whereas in the past the latter was not always relevant, with modern public participation in planning it is sometimes more difficult to get acceptance by the general public, particularly if there are well-organized and strong pressure groups in an area, than it is to get the approval of official groups. Informal liaison with local conservation groups, heritage groups, local residents groups, and bodies such as the National Trust is therefore advisable in order to determine whether support or opposition to a proposal is likely to result, and if opposition is likely, to determine what form of development might meet with approval and support.

Early discussions with the relevant development control bodies is required to determine what planning and development policies exist, to become familiar with their details, and to approach all departments concerned regarding development approval, building approval, and heritage or conservation matters in order to determine both a form of development which might receive approval, and what formal procedures must be followed with respect to the proposed project.

Matters to be researched in the feasibility study include:

(i)   land use zonings for specific land uses and development categories;
(ii)   the existence of and details of any planning guidelines;
(iii)   permitted floor/space ratios;
(iv)   requirements regarding the provision of car parking spaces or financial contributions required in lieu of provision;
(v)   the likelihood of a demand for planning contributions and the likely cost of the same where applicable.

In making such investigations, general considerations worth taking into account are that it is likely to be better to inform people of proposals at an early stage (despite the danger of your intentions becoming known to rivals) than to encounter opposition at a later stage; the approval of proposals is more likely if the development is compatible with existing neighbouring development and the locality in general; and approval of proposals is more likely if a development proposal is "environmentally friendly". With respect to the last consideration, minimization of smells and noxious fumes produced by a development, the reasonable control of traffic, the control of noise from a development, and the provision of attractive landscaping and the general enhancement of the locality are likely to improve the prospects of approval being granted.

Clarification of most planning matters at the feasibility stage may help to avoid considerable delay, unnecessary costs, and frustration at a later stage, so careful investigation of such matters is important.

Development site analysis

It is necessary to identify sites available for development and to fully investigate their suitability for the prospective project. The size of a site must be adequate, and its shape must be suitable for the project, while its physical characteristics must be investigated in respect of topography, drainage characteristics, and sub-soil conditions. The availability of suitable service provision and adequate access must also be ascertained. The asking prices for sites should be ascertained if possible, as should the other terms of sale and details of the legal interests available.

If no suitable sites can be identified the possibility of site assembly should be considered. This will necessitate researching the availability of sites, details of current ownerships, the suitability of services, problems likely to arise in acquiring interests, and the likely costs of assembly.

Details of other similar sites which may have sold should be collected for comparison purposes.

Infrastructure study

Details of current infrastructure provision should be ascertained for each site considered to be a realistic possibility for development and, if suitable provision does not already exist, the ability to provide adequate water, electricity, gas, sewage services, and telecommunications services and the likely costs of provision should be investigated, as the costs of initial provision or improvement of services may be high.

The adequacy of existing site access and the adequacy of roads contiguous to and adjacent to suitable sites must be researched with respect to the range of likely users including pedestrians, cars, service vehicles and delivery trucks, buses, and emergency services' vehicles.

When access by rail or bus is likely to be important in site selection, the adequacy of local services should be checked, as should any proposals to improve or curtail services or to alter current pricing structures, which actions might result in altered travel patterns.

A check should also be made on any regulations which would require the developer to improve any of the local infrastructure and the cost of necessary improvements or the provision of financial contributions for that purpose should be estimated for each possible site.

Survey of competition

It is extremely important to carefully check the existence of competition and the current level of competition, together with details of any proposed or possible increases in competition. Proposals to build new, competing properties in the same and nearby areas should be carefully researched, particularly those in closest proximity to the proposed development site which will probably provide the most direct competition.

If the proposed development represents a higher quality of accommodation than that currently available, for example a more prestigious type of office accommodation, the ability to upgrade and the likelihood of existing property owners upgrading existing accommodation to compete with the intended new supply should be carefully considered. The physical suitability of existing accommodation for upgrading should be considered, as should the costs and the time required for doing so. If upgrading of existing accommodation appears feasible, it should not be overlooked that such upgraded space could possibly be brought to the market before a new development, nor should it be overlooked that such a possibility might provide a more realistic development opportunity for the developer than developing completely new property.

Useful sources of information regarding competition are likely to include local planning authorities, real estate agents, the local press, local contacts, the business press, and the property press in particular.

The critical thing to be determined is the level and quality of competition which is likely to exist on completion of the proposed development project.

Financial feasibility study

Completion of the previous stages of the overall feasibility study should provide enough information for a reasonably exacting financial appraisal of the proposed project to be made.

The appraisal process is described in Chapters 13 and 14. As observed in those chapters the valuer will have to make judgements based on the information collected through research and investigation, and will also have to make forecasts regarding the future which will be based on the information collected, personal judgements, and personal expectations.

As already emphasised and as indicated in the examples in this book, there is great scope for widely varying valuation results to be produced from the same basic information. The valuer should therefore attempt as early as possible to stabilise as many of the inputs as possible by methods indicated earlier in this book and, in the interim, any initial appraisals should be treated with caution and account should be taken of sensitivity analyses.

Funding arrangements

Developments can only be undertaken if sufficient funds are available, and they will only be successful in financial terms if finance is arranged on suitable terms. The feasibility study should therefore research possible sources of finance determining the amounts each source would be prepared to lend and the terms (particularly the interest terms) on which funds would be advanced. The level of arrangement fees and servicing fees should be determined as should the existence of penalty clauses for late payment of sums due or for extension of the loan period. The requirements for security should also be ascertained.

Enquiries should relate to both short-term funding, which is particularly relevant to the development project, and long-term funding which may well influence the ability to sell the completed development and the figure at which it may sell.

Financial information should be obtained before too much detailed research work is undertaken, as unless suitable funding is likely to be available a development is unlikely either to be possible or to be profitable, and, if neither is likely, further investigatory work will be futile.

Summary of feasibility study objectives

A developer's objective is to successfully complete an envisaged development project, success including making an acceptable profit. The objective of the feasibility study is to indicate whether financial success is likely to result from a project and what level of success is likely, and a high-quality study should be undertaken to involve full investigation of all the factors relevant to the likely success or otherwise of a development project. Such a study might include all or some of the items contained in the above summary, and it might be necessary to research other matters not contained therein.

As with financial appraisals, it is arguable that the most important feasibility studies are those that reveal that a project is likely to be unprofitable or that the likely profit is too small to justify undertaking the project and accepting the inherent risks.

Chapter 19

Concluding Observations

Consideration of the range of matters suggested for inclusion in a feasibilty study should remind the reader just how complex an undertaking a property development project can be. Indeed, even small-scale projects provide a range of challenges for a developer, and there is unlikely to be any project which is not subject to a considerable element of risk and uncertainty.

There is therefore always the possibility of financial loss rather than financial gain, a fact which should be remembered by those who sometimes begrudge successful property developers their success. The profit motive is an important stimulus to development and, where profitability is unlikely to exist, development is in most circumstances unlikely to be undertaken. The provision of suitable properties for the range of activities undertaken by society in general is an important service to society, be it through new development, redevelopment, or refurbishment of existing properties and, without adequate and timely provision of suitable properties, an economy is likely to function less efficiently and less effectively, while at least some sectors of society are likely to be disadvantaged. One has only to think of the plight of those living in poor residential conditions or those who are homeless to appreciate the truth of the latter statement.

In order to develop property successfully, and to thereby serve society more effectively, it is essential for those involved in property development to understand fully the range of problems likely to be encountered in any project, and to undertake exhaustive research to ensure that they are fully informed before committing themselves to a project. It is also essential that they address all identified problems, and also that they consider what other problems might occur, in order that exposure to risk can be mitigated and avoided if possible. It is extremely important that a developer retains control of a development project from its first inception, as when circumstances control a developer (rather than the reverse) problems are likely to arise which may, with the passage of time, prove to be insurmountable.

It should not be forgotten that failure in property development is not a remote and unlikely possibility, as consideration of the terrible problems experienced by many property developers in the late 1980s and early 1990s in many countries throughout the world is evidence to. Long established, well-staffed, experienced, and apparently financially sound property development companies experienced enormous problems to the extent that many of them became bankrupt, their problems sending substantial ripples through the economies of many countries with resultant adverse effects on many other economic activities, particularly those requiring substantial bank funding.

No matter how competent an organization is and no matter how exhaustive and thorough initial research may be, no one can control the future and eliminate uncertainty, and even the ". . . best laid plans of mice and men . . ." can be thwarted when circumstances move against them in a way which they cannot control.

Olympia and York, the Canadian property developers, were in the 1980s the largest property development company in the world, with great experience and enormous resources and expertise at their disposal. Within a relatively short period of time they became insolvent, brought down primarily by financial problems with their extremely large Canary Wharf undertaking in London's Docklands. There is little doubt that their involvement was only undertaken after considerable research and thought, yet possibly two key factors brought about their eventual downfall. The first was probably the inadequacy of the service infrastructure, in particular the inadequacy of rail links, to the Docklands area. This was a factor in the overall disappointing economic performance of the area which it could be argued should have been foreseen by the company. However, these problems were compounded by the severe economic recession of the late 1980s and early 1990s from which the company was unable to recover; the overall effect on the economy being greater than the vast majority of commentators ever envisaged, even if they had been perceptive enough to forecast bad times ahead.

In 1996 The Myer Centre in Adelaide, South Australia was sold for the reported figure of Australian $151 million. It had been developed on an estimated cost reported as about $350 million, which in fact turned out to be about $500 million on completion of the project, and commentators estimated that the development costs plus accumulated interest charges probably totalled about

$900 million by the date of sale. Such a major project must have been extensively researched before it was undertaken, and many experienced professionals, including financial specialists, must have approved the project for it to be commenced, yet success was not assured and with the passage of time the development became a major financial failure.

These are just two examples of major projects which went wrong and they were not isolated examples during that period; a very large list of failed projects and failed companies could be compiled if necessary to prove the point. Property development and risk and uncertainty cannot be completely divorced from each other, no matter how meticulous research and how skilful project management may be, so risk identification and control must be major activities for property developers if a high level of success is to be achieved with projects.

The author has attempted to discuss property development taking into account not only the considerations of property developers and members of the supporting development teams, but also society in general. Many parties may be affected by any one development, whilst society in general ultimately determines the utility of and therefore the value of developed properties. Societal considerations are therefore extremely important.

Emphasis has been given to the importance of the development valuation in the decision making process and of the need to fully research all relevant matters and all inputs to the valuation process. The author believes that it is probably more important to ask the right questions than it is to believe one has the right answers to every question, for it will be impossible to determine whether there is a right answer in some cases in which a range of possibilities may exist, and he would not profess to have necessarily answered all the questions raised in this book. Indeed, it is probably extremely dangerous in the property development process to believe one has all the right answers, and continual review of the continuing process is an essential part of the developer's role. It is important always to keep an open mind and to remember that in most situations there are likely to be a range of acceptable solutions or courses of action, while changing circumstances may produce changing needs and result in new courses of action and new solutions becoming appropriate.

It is important to look at all aspects of problems and to consider viewpoints other than one's own if one is to adequately understand

and consider problems and alternative approaches to situations. One of the objectives in this book has been to attempt to consider a range of views and possible approaches to a variety of considerations relevant to property development. It is hoped that the content has been of benefit to the reader and details of further books which contain relevant information are provided on the following page.

Further Reading

The following books cover various aspects of the property development process and are recommended by the author for further reading:

Bailey, A., *How to be a Property Developer*, Mercury Books, London, 1988.

Britton, W., Davies, K. & Johnson, T. A., *Modern Methods of Valuation*, The Estates Gazette Limited, London, 1989.

Darlow, C. (Editor), *Valuation and Development Appraisal*, The Estates Gazette Limited, London, 1988. This book is out of print but is available in libraries.

Harvey, J., *Urban Land Economics*, MacMillan, London, 1996.

Miles, M.E., Malizia, E.E., Weiss, M.A., Berens, G.L., & Travis, G., *Real Estate Development Principles and Process*, The Urban Land Institute, Washington D.C., 1994.

Rees, W.H. (Editor), *Valuation: Principles into Practice*, The Estates Gazette Limited, London, 1988.

Rose, Jack, *The Dynamics of Urban Property Development*, E. & F.N. Spon, London, 1985.

Topping, R. and Avis, M. (Editors), *Property Development*, E. & F.N. Spon, London, 1995.

Index